De Volson Wood

Treatise on the Theory of the Construction of Bridges and Roofs

De Volson Wood

Treatise on the Theory of the Construction of Bridges and Roofs

ISBN/EAN: 9783337219505

Printed in Europe, USA, Canada, Australia, Japan

Cover: Foto ©berggeist007 / pixelio.de

More available books at **www.hansebooks.com**

TREATISE

ON THE

THEORY OF THE CONSTRUCTION

OF

BRIDGES AND ROOFS.

ILLUSTRATED WITH NUMEROUS WOOD ENGRAVINGS.

BY

DE VOLSON WOOD,

Professor of Mathematics and Mechanics in Stevens' Institute of Technology, Formerly Professor of Civil Engineering in the University of Michigan.

THIRD EDITION
REVISED AND CORRECTED.

NEW YORK:
JOHN WILEY & SONS, PUBLISHERS,
15 ASTOR PLACE.
1883.

Entered according to Act of Congress, in the year 1875,
By DE VOLSON WOOD,
In the Office of the Librarian of Congress, at Washington.

TABLE OF CONTENTS.

PART I.

FORMULAS PERTAINING TO BEAMS.

NO. OF THE ARTICLE.		PAGE
1.	Introductory Remark	1
2.	Notation	2
3.	Case of a Horizontal Beam which is fixed at one End and loaded at the free End	2
4.	Case of a Horizontal Beam which is fixed at one End, and loaded uniformly	3
5.	Case of a Horizontal Beam which is supported at its Ends, and loaded at any point	5
6.	Case of a Horizontal Beam which is supported at its Ends and loaded at the middle point	6
7.	Case of a Beam which is supported at its Ends and loaded uniformly	7
8.	Case of a Beam which is fixed at one End, supported at the other, and loaded at any point	8
9.	Case of a Beam which is fixed at one End, supported at the other, and loaded uniformly	9
10.	Case of a Beam which is fixed at both Ends and loaded at the middle point	10
11.	Case of a Beam which is fixed at both Ends, and loaded uniformly	10
12.	Case of a Uniform Load and uniformly-increasing Load	11
13.	Partial Uniform Load	12
14.	Case of a Load applied at several points	13
15.	Case of Oblique Strains	14
16.	Inclined Beam loaded at one point	15
17.	A Loaded Brace	19
18.	Discussion of the preceding case	20
19.	Problems	21
20.	Rafter loaded uniformly	21
21.	Strength of Pillars	23
22.	Gordon's Rules	24
23.	Mr. C. Shaler Smith's Formula	25
24.	Hodgkinson's Formulas	26
25.	Weight of Pillars	27

PART II.

TRUSSED BRIDGES.

CHAPTER I.

DEFINITIONS AND DATA.

NO. OF THE ARTICLE.		PAGE
26.	Definitions	28
27.	The Load	31
28.	Safe Load	33
29.	Factor of Safety	33
30.	Absolute Modulus of Safety	35
31.	Examples of Strains that have been used in Practical Cases	35
32.	Proof Load	39
33.	Framing	39
34.	The Simple Girder	39

CHAPTER II.

KING-POST SYSTEMS.

35.	The King-post Truss	43
36.	Inverted King-post	47
37.	A Braced Beam	47
38.	Minimum Amount of Material	47
39.	Minimum Depression	50
40.	Trussed Beams	51
41.	Raised Tie or Double Rafters	52
42.	Depressed Tie	55
43.	Solution by Diagrams	55
44.	Fink Truss	56
45.	Roof Trusses	58
46.	Unequally-inclined Braces or Ties	58
47.	Inverted Truss	60
48.	Bollman's Truss	61
49.	Analysis of Bollman's Truss Primary System	63

CHAPTER III.

TRUSSES WITH PARALLEL CHORDS.

50.	Trapezoidal Truss	66
51.	Truss Fully Braced	67
52.	Trapezoidal Truss Modified	68

TABLE OF CONTENTS.

NO. OF THE ARTICLE.	PAGE
53. Strains upon the Modified Truss	68
54. Trapezoidal Truss Inverted	68
55. Examples	68
56. Triangular Truss—Definition	70
57. Triangles the only proper Figures for Trusswork	70
58. Conditions Assumed	71
59. Notation	71
60. CASE I	72
61. Strains on the Ties and Braces	73
62. Strains on the Chords	73
63. CASE II	74
64. Strains on the Ties and Braces	74
65. Strains on the Chords	75
66. Open Swing Bridges	76
67. CASE III	77
68. Geometrical Solution	77
69. Observations on the Preceding Results	78
70. Distribution of Strains on the Ties	78
71. Observations on the Preceding Analysis	80
72. Formulas for Strains on the Tie-Braces for a Uniform Load	82
73. Another Solution of the Preceding Problem	85
74. Maximum Strains on the Tie-Braces	85
75. Strains on the Chords	86
76. Sub-Case; in which the Number of Bays in the Lower Chord is odd	89
77. Inverted Truss	90
78. Weight of the Truss Considered	90
79. Discussion of the Equation (111)	92
80. Problem	94
81. Strains on the Chords when the Weight of the Truss is considered	94
82. Dimensions of the Ties	94
83. Dimensions of the Braces	95
84. Dimensions of the Chords	95
85. CASE IV.—Distribution of the Strains	96
86. Stress on the Tie-Braces	97
87. General Method	99
88. Problems	99
89. Strains on the Chords	101
90. Problem	102
91. Examples	102
92. Minimum Amount of Material	103
93. Proper Length of the Bays	106
94. CASE V	106
95. Stress on the Braces	106
96. Stress on the Chords	107
97. Towne's Lattice	107
98. Analysis of Towne's Lattice	108
99. Analysis of the Multiple Triangular System	109

TABLE OF CONTENTS.

NO. OF THE ARTICLE.		PAGE
100.	Ambiguity in regard to Strains in certain Cases	111
101.	Warren's Girder Modified	112
102.	Type Form of Howe's Truss	114
103.	Type Form of Pratt's Truss	114
104.	Definition of the Panel System	115
105.	Maximum Stress on the Diagonals	115
106.	Discussion of Equation (128)	117
107.	A Counter-Brace	120
108.	Values of N_0	121
109.	Examples	121
110.	General Value of the Second Differences	123
111.	Stress on any Diagonal	124
112.	Uniformly Distributed Load	125
113.	Stress on the Verticals	127
114.	Stress upon the Chords	129
115.	Examples	130
116.	Load concentrated at one Point	131
117.	Stress on the Chords for an unequally-distributed Load	132
118.	Multiple Systems	133
119.	Problem	133
120.	A Second Problem of Minimum Material	136
121.	General Problem of Minimum Material	136
122.	Minimum Material in a Post and Tie combined	140
123.	Long's Truss	143
124.	Howe's Truss	145
125.	Main Braces in Howe's Truss	146
126.	Counter-Braces	147
127.	Keying the Counter-Braces	148
128.	Cambering	149
129.	The Block in Howe's Truss	149
130.	The Vertical Tie-Rods in Howe's Truss	150
131.	The Chords in Howe's Truss	150
132.	Pratt's Truss	150
133.	Whipple's Truss	151
134.	Analysis of the Double-Panel System as shown in Fig. 92	155
135.	Post's Truss	162
136.	Extra Strains considered	170
137.	Haupt's Lattice	170
138.	Hall's Lattice	171
139.	Lateral Bracing	171
140.	Knee-Braces	172
141.	Stability of the Bridge upon its Supports	172
142.	Continuous Loading—Vertical Shearing Stress	173
143.	General Problems Unsolved	176
144.	Law of Strains upon the Chords	177
145.	Law of Relation between the vertical Shearing Stress and Moments of Applied Forces	178

CHAPTER IV.

TRUSSES WITH CHORDS NOT PARALLEL.

No. of the Article.		Page
146.	McCallum's Truss	179
147.	PARABOLIC ARCHED TRUSS—Definition and Notation	180
148.	Case of Uniform Load	182
149.	Case of a Partial Uniform Load—when the Diagonals are Ties	183
150.	Case of a Partial Uniform Load—when the Diagonals are Braces	184
151.	Triangular Trussing—Parabolic Arched Truss	185
152.	Strains on the upper Chord—found by Moments	186
153.	A General Problem	187
154.	Both Chords Curved	188

CHAPTER V.

COMPOUND STRUCTURES.

155.	Definition	190
156.	Burr Truss	191
157.	Pennsylvania Rail-Road Bridge	191
158.	A Third Example of Compound Structures	192

PART III.

ROOFS.

159.	Definitions	193
160.	Roof Trusses	193
161.	General Data	194
162.	Description of the Roof over the Large Hall at the University of Michigan	194
163.	Load on the Flat Part of the University Roof	197
164.	Weight of Snow	199
165.	The Pressure of Wind	199
166.	Weight of the Dome on the University Roof	201
167.	Weight of the Main Trusses	201
168.	Weight of the Cross Truss	201
169.	Results Collected	201
170.	Analysis of the University Roof Truss	202
171.	Cambre of Trusses	203
172.	Case of Secondary Trussing in which the Bays in the Lower Chord are equal to each other	211
173.	Another Form of Roof Trussing	218
174.	The preceding Form modified	223

PART IV.

GENERAL PROBLEM OF TRUSSED GIRDERS.

NO. OF THE ARTICLE.	PAGE
175. GENERAL EQUATIONS.	224
176. Forces in a Plane.	226
177. Applied Forces Vertical.	226
178. Lower Chord Horizontal.	231
179. Upper Chord Horizontal.	232
180. Both Chords Horizontal.	232
181. Case of a Horizontal Beam.	234
182. PERFECTLY FLEXIBLE SYSTEM.	235
183. Forces in a Plane.	237
184. An Inverse Problem.	241

APPENDIX I.

DEMONSTRATION OF A GRAPHICAL MODE OF SOLVING PROBLEMS.

APPENDIX II.

TABLE OF THE MECHANICAL PROPERTIES OF THE MATERIALS OF CONSTRUCTION.

PREFACE.

THIS book contains the substance of my lectures upon Trussed Bridges and Roofs as given to the senior class in civil engineering in the University of Michigan. They were mostly delivered extempore, but in preparing them for publication I have followed the same order and treated of the same subjects as were given to the class, excepting that in the lectures the problems pertaining to "Minimum Material" were omitted.

The lectures were intended to give to the student a correct knowledge of the elements of the subject. There was but a slight attempt to treat of the details of the several trusses which were considered. Indeed it is generally unprofitable to treat of isolated details in the class-room. As a general rule, only such subjects should be taught in technical courses as admit of classified principles. The conditions which determine all the qualities of a "detail" are so infinitely varied that only *general* rules can be given for their construction. I am aware, however, that much more can profitably be done in this direction than has been here attempted.

During the lectures I often suggested problems as exercises for the students. Some of these I have entered in the body of the work, and I have with pleasure given credit to those students who first solved them.

I have not been able to prepare in time for this book the lectures upon Tubular and Suspension Bridges and Arches.

De V. W.

Hoboken, N. J., Jan. 11th 1872.

TREATISE

ON

BRIDGES AND ROOFS.

Part 1.

FORMULAS PERTAINING TO STRAINS ON SINGLE PIECES.

1.—INTRODUCTORY REMARK.—In proportioning bridges and roofs, and other similar mechanical structures, we generally assume, at first, that the several parts are reduced to physical rigid lines, and the several stresses to which they are to be subjected are determined on statical principles, without regard to the sizes of the pieces which are to be used. After the stresses have been determined, the several pieces are proportioned to resist those stresses. The size of each piece may thus be determined separately, without any reference to the size or position of the other pieces which compose the structure. We may therefore determine the formulas which are applicable to single pieces when they are subjected to different strains, without regard to the office which they are to perform in a given structure. This has been done for several of the more simple cases in the author's *Resistance of Materials*.

Several of the formulas which are applicable to these cases are here brought together for convenience in use; for the proof of which the reader is referred to the author's work above mentioned. Some other cases are also added with their accompanying demonstrations.

2.—NOTATION.—The following notation is used, which is here arranged alphabetically.

A = the area of a transverse section of a piece.
b = the breadth of a rectangular piece.
C = the *modulus of resistance* to crushing.
d = the depth of a rectangular piece.
δ = the weight of a cubic inch of volume.
Δ = the maximum deflection of a piece which is bent.
d_1 = the distance of the most remote fibre from the neutral axis in a bent piece.
D = the external diameter of a cylinder.
D_1 = the internal diameter of a hollow cylinder.
E = the *coefficient of elasticity* for a longitudinal stress.
E_s = the *coefficient of elasticity for transverse shearing stress*, the approximate value of which is = $\frac{1}{5} E$.
I = the moment of inertia of a section.
l = the length of a beam, column, or other simple pieces.
M = the maximum moment of applied forces.
M_x = the general moment of applied forces.
P = the load applied at a single point of a beam or other piece.
r = the radius of a circle, and when it is external, let
r_1 = the radius of the internal circle.
R = the *modulus of rupture* in a bent beam.
S = the *modulus of rupture* for transverse shearing stress.
S_s = the total transverse shearing stress.
T = the *modulus of tenacity*.
w = the load per unit of length when it is uniformly distributed.
W = the total load on a beam.
x = the variable abscissa.
y = an ordinate perpendicular to the axis of the piece.

For values of the *moduli* and coefficients, see Appendix.

3.—CASE OF A HORIZONTAL BEAM WHICH IS FIXED AT ONE EXTREMITY AND HAS A WEIGHT, P, RESTING UPON THE FREE EXTREMITY; as in Fig. 1.

The general moment of P is

$$Px.$$

FIG. 1.

The maximum moment of P is
$$Pl.$$
The general value of the transverse shearing stress is *
$$S_s = P.$$
The dimensions of the beam to resist rupture may be found from the formula
$$Pl = \frac{RI}{d_1},$$ and if the beam is rectangular we have
$$Pl = \tfrac{1}{6} Rbd^2 \dots\dots\dots\dots (1)$$
If the beam is rectangular the maximum deflection is
$$\varDelta = 4\,\frac{Pl^3}{Ebd^3} + \frac{Pl}{E_s bd}$$
When the beam is long compared with the depth, we have with sufficient accuracy
$$\varDelta = 4\,\frac{Pl^3}{Ebd^3} \dots\dots\dots\dots (2)$$

EXAMPLE.—A beam whose length is 12 feet is fixed at one end, free at the other, $b = 4$ inches, $d = 12$ inches, $E = 1,000,000$ pounds. Required the weight at the free end which will deflect it two inches.†

4.—SUPPOSE THAT THE BEAM IS FIXED HORIZONTALLY AT ONE END, IS FREE AT THE OTHER, AND HAS A LOAD UNIFORMLY DISTRIBUTED OVER ITS WHOLE LENGTH.— The beam may be fixed as shown in Figs. 2 and 3.

The general moment of the load is
$$\tfrac{1}{2} w x^2.$$

---

* The general value of the transverse shearing stress is the first differential coefficient of the general moment of applied forces in reference to x.

† When used as a text-book the student should be required to solve the examples. The answers are purposely omitted.

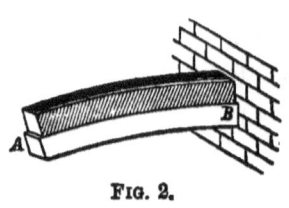
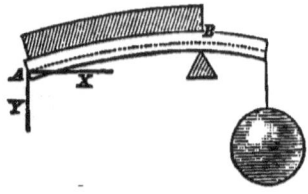

FIG. 2.   FIG. 3.

The maximum moment of the load is

$$\tfrac{1}{2} Wl.$$

The general value of the transverse shearing stress is

$$S_s = wx.$$

The maximum shearing stress is

$$S_s = wl = W.$$

The dimensions of the beam to resist rupture from transverse strain may be found from the formula

$$\tfrac{1}{2} Wl = \frac{RI}{d_1};$$

and if the beam is rectangular we have

$$\tfrac{1}{2} Wl = \tfrac{1}{6} Rbd^2 \dots\dots\dots\dots\dots\dots (3)$$

If the beam is prismatic and rectangular, the maximum deflection is

$$\Delta = \tfrac{3}{2} \frac{Wl^3}{Ebd^3} + \frac{Wl}{2E_sbd};$$

and if the beam is long compared with the depth, we have with sufficient accuracy

$$\Delta = \tfrac{3}{2} \frac{Wl^3}{Ebd^3} \dots\dots\dots\dots\dots (4)$$

EXAMPLE.—If $E = 1,200,000$ pounds, $E_s = 400,000$ pounds, $b = 3$ inches, $d = 6$ inches, $l = 5$ feet, and the load 20,000 pounds uniformly distributed over the whole length, required the maximum deflection. Also the value of R in Eq. (3.)

5.—LET THE BEAM BE HORIZONTAL, SUPPORTED AT ITS ENDS AND A WEIGHT APPLIED AT ANY POINT.

Figs. 4 and 5 represent the case.

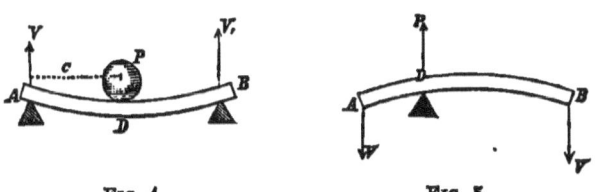

FIG. 4.   FIG. 5.

Let $c = AD$, the distance of P from A, then the general moment of P between A and D is

$$\frac{l-c}{l} Px \dots \dots (5)$$

and between D and B it is

$$Pc\frac{l-x}{l} \dots \dots (6)$$

The maximum moment of the load is at D, and is

$$\frac{l-c}{l} cP \dots \dots (7)$$

The general value of the transverse shearing stress between A and D is

$$\frac{l-c}{l}P,$$

and between D and B it is

$$-\frac{c}{l}P.$$

If the beam is rectangular, its dimensions to resist rupture may be determined from the formula

$$\frac{l-c}{l} cP = \tfrac{1}{6} Rbd^2.$$

The deflection of the beam at D is (omitting the shearing resistance)

$$\frac{P c^3}{3EIl}(c-l)^2.$$

If c exceeds $\tfrac{1}{2} l$, the maximum deflection will be between A and D; but if it is less it will be between D and B. In all cases c may be taken on the end which exceeds $\tfrac{1}{2} l$. If it is so taken, the point of maximum deflection is

$$x = \sqrt{\frac{c^3 + 2cl^2 - 3c^2 l}{3(l-c)}}.$$

This value, substituted in the following expression:

$$y = \frac{P}{6EIl}\left[(c-l)x^3 + (c^3 + 2l^2 - 3cl)cx\right]$$

will give the maximum deflection.

The expression thus becomes very complicated. The more common case is that in which the load is at the middle of the beam.

6.—LET THE BEAM BE HORIZONTAL, SUPPORTED AT ITS ENDS, AND SUSTAIN A LOAD, P, placed at the *middle*.

The general moment of P is

$$\tfrac{1}{2} Px. \quad\quad\quad\quad\quad (8)$$

The maximum moment is at the middle of the beam, and is

$$\tfrac{1}{4} Pl. \quad\quad\quad\quad\quad (9)$$

The general value of the transverse shearing stress is

$$\tfrac{1}{2} P. \quad\quad\quad\quad\quad (10)$$

If the beam is prismatic and rectangular, we have for the dangerous section

$$\tfrac{1}{4} Pl = \tfrac{1}{6} Rbd^2. \quad\quad\quad\quad\quad (11)$$

The maximum deflection of the beam is at the centre, and is

$$\varDelta = \frac{Pl^3}{48EI} + \frac{Pl}{4E_sA}, \quad\quad\quad\quad\quad (12)$$

which becomes for rectangular beams

$$\varDelta = \frac{Pl^3}{4Ebd^3} + \frac{Pl}{4E_sbd}. \quad\quad\quad\quad\quad (13)$$

and, by omitting the last term, becomes

$$\varDelta = \frac{Pl^3}{4Ebd^3}. \quad\quad\quad\quad\quad (14)$$

EXAMPLES. 1.—A beam whose length is 15 feet, is supported at its ends, and is to sustain a weight, $P = 3,000$ pounds, placed at the middle. Required the breadth and depth, so that d shall be $4b$, and $R = 1,000$ pounds.

2.—If $l = 12$ feet, $b = 1$ inch, $d = 4$ inches, $E = 27,000,000$ pounds, $E_s = \frac{1}{6} E$, how much load placed at the middle will be required to deflect the beam $\frac{1}{4}$ of an inch?

7.—LET THE BEAM BE HORIZONTAL AND SUPPORTED AT OR NEAR ITS EXTREMITIES, AND HAVE A LOAD UNIFORMLY DISTRIBUTED OVER ITS WHOLE LENGTH.

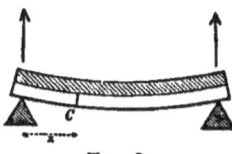

FIG. 6.

The general moment of the load is
$$\tfrac{1}{2} w (lx - x^2) \quad \ldots \ldots \ldots \ldots (15)$$
The maximum moment is at the middle of the beam, and is
$$\tfrac{1}{8} wl^2 = \tfrac{1}{8} Wl \quad \ldots \ldots \ldots \ldots (16)$$
The general value of the transverse shearing stress is
$$\tfrac{1}{2} wl - wx \quad \ldots \ldots \ldots \ldots (17)$$
The maximum shearing stress is at the ends, and may be found by making $x = 0$ or $x = l$ in the preceding expression; hence its value is
$$\tfrac{1}{2} wl \text{ or } -\tfrac{1}{2} wl = \pm \tfrac{1}{2} W \ldots (18)$$
If the beam is prismatic and rectangular, we have for the dangerous section
$$\tfrac{1}{8} Wl = \tfrac{1}{6} Rbd^2 \quad \ldots \ldots \ldots (19)$$
The maximum deflection is
$$\Delta = \frac{5}{384} \frac{wl^4}{EI} + \frac{wl^2}{8 E_s A}, \quad \ldots \ldots (20)$$
which for rectangular beams becomes
$$\Delta = \frac{5}{32} \frac{Wl^3}{Ebd^3} + \frac{Wl}{8 E_s bd}, \quad \ldots \ldots (21)$$
and by omitting the last term becomes
$$\Delta = \frac{5}{32} \frac{Wl^3}{Ebd^3} \quad \ldots \ldots \ldots \ldots (22)$$

EXAMPLE.—If a beam whose length, $l = 10$ feet, depth, $d = 3$ inches, breadth, $b = \frac{1}{4}$ inch, coefficient of elasticity, $E = 25,000,000$ pounds, is supported at its ends and uniformly loaded; required the deflection when the greatest strain on the fibres is 12,000 pounds per square inch. Use equations (19) and (22).

8.—LET THE BEAM BE HORIZONTAL, FIXED AT ONE EXTREMITY, SUPPORTED AT THE OTHER, AND HAVE A WEIGHT, P, APPLIED AT ANY POINT. Figs. 7, 8, and 9.

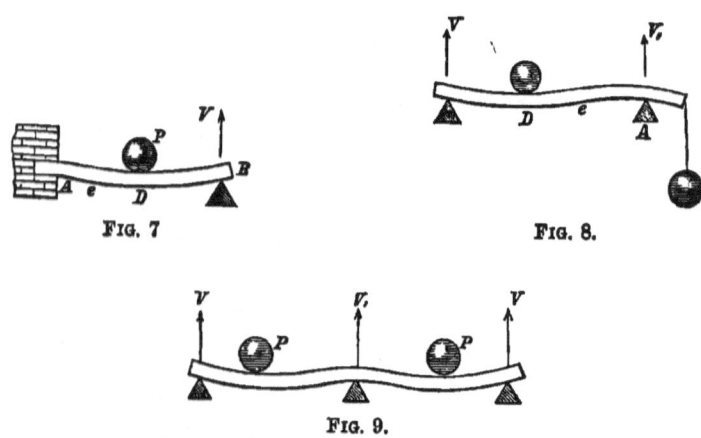

FIG. 7.   FIG. 8.

FIG. 9.

To produce the greatest strain, P must be placed at a distance of 0.634 of the length of the beam from A; or $AD = 0.634\,l$. For this case the general moment of the load on DB is

$$-0.475\,(l-x)\,P,$$

and for the part DA it is

$$(0.159\,l - 0.525\,x)\,P.$$

The maximum moment of stress is

$$0.174\,Pl.$$

The general value of the transverse shearing stress is

$$S_s = 0.475\,P \text{ for } DB, \text{ and}$$
$$= 0.525\,P \text{ for } DA.$$

For prismatic rectangular beams we have for the dangerous section

$$0.174\, Pl = \tfrac{1}{6} Rbd^2 \quad \ldots \ldots \ldots \ldots (23)$$

The maximum deflection (omitting shearing resistance) is at $0.6045\, l$ from A, and is

$$\varDelta = 0.00957\, \frac{Pl^3}{EI} \quad \ldots \ldots \ldots \ldots (24)$$

For rectangular beams, $I = \tfrac{1}{12} bd^3$.
For cylindrical beams, $I = \tfrac{1}{4} \pi r^4$.

9.— LET THE BEAM BE HORIZONTAL, FIXED AT ONE END, SUPPORTED AT THE OTHER, AND UNIFORMLY LOADED OVER ITS WHOLE LENGTH.

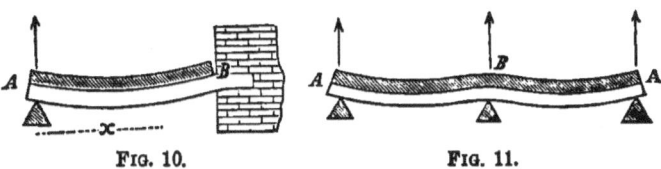

FIG. 10.   FIG. 11.

The general moment of stress is

$$M_x = \tfrac{1}{8} wx(4x - 3l) \quad \ldots \ldots (25)$$

The maximum moment of stress is

$$M = \tfrac{9}{128} Wl \quad \ldots \ldots \ldots \ldots (26)$$

The general value of the transverse shearing stress is

$$S_s = wx - \tfrac{3}{8} wl.$$

The maximum shearing stress is at B, and equals

$$S_s = \tfrac{5}{8} wl = \tfrac{5}{8} W.$$

For the dangerous section, for prismatic rectangular beams we have

$$\tfrac{1}{8} Wl = \tfrac{1}{6} Rbd^2 \quad \ldots \ldots \ldots \ldots (27)$$

The maximum deflection is

$$\varDelta = \tfrac{1}{185} \frac{Wl^3}{EI} \quad \ldots \ldots \ldots \ldots (28)$$

QUESTIONS.—How does the maximum deflection in this case compare with that in Article 7 for the same load? In which case is the beam most liable to break? In which case is the shearing stress greater at the middle?

10.—LET THE BEAM BE HORIZONTAL, FIXED AT BOTH ENDS, AND A WEIGHT REST UPON IT AT THE MIDDLE POINT.

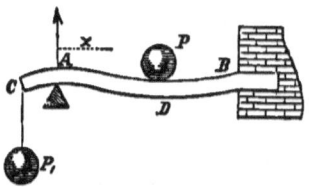

Fig. 12.

The general moment of stress is
$$M_x = \tfrac{1}{8} P(l-4x).$$

The maximum moment is at A, or B, or D, being the same for each point, and is
$$M = \pm \tfrac{1}{8} Pl \dots \dots \dots (29)$$

The general value of the shearing stress is
$$Ss = \tfrac{1}{2} P.$$

The dangerous section for prismatic beams is at A, or B, or D, and for rectangular sections, we have
$$\tfrac{1}{8} Pl = \tfrac{1}{6} Rbd^2 \dots \dots \dots (30)$$

The maximum deflection is (omitting shearing stress)
$$\frac{Pl^3}{192\, EI} \dots \dots \dots (31)$$

11.—LET THE BEAM BE HORIZONTAL, FIXED AT BOTH ENDS, AND A LOAD UNIFORMLY DISTRIBUTED OVER ITS WHOLE LENGTH.

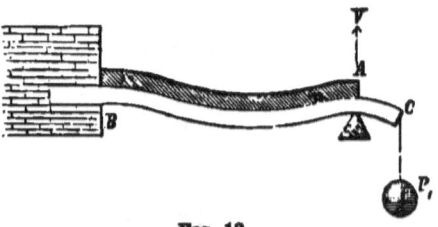

Fig. 13.

The general moment of stress is
$$M_x = \frac{W}{12l}\left[l^2 - 6x(l-x)\right] \dots (32)$$

The maximum moments are

$M = \frac{1}{12} Wl$, which is for A and B .... (33)
$M = \frac{1}{24} Wl$, which is for the middle of the beam.

The general value of the transverse shearing stress is

$$S_e = \tfrac{1}{2} wl - wx.$$

The maximum value of the transverse shearing stress, which is at the ends, is

$$S_e = \pm \tfrac{1}{2} wl.$$

At the dangerous section for prismatic rectangular beams, we have

$$\tfrac{1}{12} Wl = \tfrac{1}{6} Rbd^2 \ldots\ldots\ldots (34)$$

The maximum deflection is (omitting shearing stress)

$$\Delta = \tfrac{1}{384} \frac{Wl^3}{EI} \ldots\ldots\ldots\ldots (35)$$

12.—INCREASING LOAD.—LET THE BEAM BE HORIZONTAL, SUPPORTED AT ITS END, AND LOADED UNIFORMLY OVER ITS WHOLE LENGTH, AND ALSO AN ADDITIONAL LOAD WHICH BEGINS WITH NOTHING AT ONE END AND INCREASES UNIFORMLY TO THE OTHER, as in Fig. 14. The

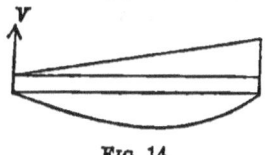

FIG. 14.

space between the lower convex line and the horizontal one just above it, is the form of the beam of uniform strength for this case.

Let $W =$ the total weight of the uniform load,
$W_1 =$ the total weight of the uniformly increasing load,
and the other notation as before.

The general moment of stress is

$$M_x = (\tfrac{1}{2} W + \tfrac{1}{6} W_1) x - \frac{Wx^2}{2l} - \frac{W_1 x^3}{3l^2}$$

The maximum moment of stress is at the point

$$x = \left[-\tfrac{1}{2} W \pm \sqrt{\frac{W^2}{4} + (\tfrac{1}{2}W + \tfrac{1}{6}W_1)\, W_1} \right] \frac{l}{W_1},$$

which value, substituted in the preceding equation, gives the maximum moment. The case becomes much simplified by neglecting the uniform load. Letting $W = 0$, and the general moment becomes

$$M_x = \tfrac{1}{8} W_1 x - \frac{W_1 x^2}{3l^2}$$

and the abscissa of the point of maximum stress becomes $x = \tfrac{1}{8} \sqrt{3l}$, which in the preceding equation gives the maximum moment,

$$M = \tfrac{2}{27} \sqrt{3} \ W_1 l.$$

The general expression for the transverse shearing stress, for the general case, is

$$S_s = \tfrac{1}{2} W + \tfrac{1}{8} W_1 - \frac{Wx}{l} - \frac{W_1 x^2}{l^2}.$$

The point of maximum shearing stress is for $x = l$, and its value is

$$S_s = + \tfrac{1}{2} W + \tfrac{3}{8} W_1.$$

13.—PARTIAL LOAD. LET THE BEAM BE HORIZONTAL, SUPPORTED AT ITS ENDS, AND A UNIFORM LOAD OVER A PORTION OF ITS LENGTH. Fig. 15.

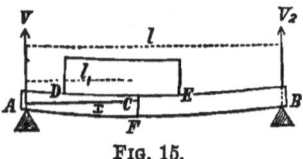

FIG. 15.

Let $2a = DE =$ the length of the uniform load;
$x = AF =$ the distance to any section from A;
C the centre of the load;
$l_1 = AC$; and
$l_2 = BC.$

There are three cases, as follows:
The general moment between A and D is

$$M_x = Vx \text{ or } 2wa \frac{l_2}{l} x,$$

in which x must not exceed AD.

The general moment of stress under the load is
$$M_x = Vx - \tfrac{1}{2}w(x - l_1 + a)^2,$$
$$= \frac{2wal_2}{l}x - \tfrac{1}{2}w(x - l_1 + a)^2.$$

The maximum moment of stress is

$$M = \frac{wa}{l^2}\left[2l_2[l(a + 2l_1) - 2al_1] - (l - l_1)\right]\ldots\ldots(36)$$

The general moment of stress between B and E is
$$M_x = -V_2(l-x),$$
in which x must equal or exceed the horizontal distance AE.

The shearing stress from A to D is
$$S_s = V.$$

The shearing stress under the load, from D to E, is
$$S_s = V - w(x - l_1 + a)\ldots\ldots(37)$$

The shearing stress from E to B is
$$S_s = -V_2 = -2wa\frac{l_1}{l}$$

PROBLEMS.—Find the value of M Eq. (36) when $l_1 = a$, and discuss it in reference to l_2. Find the value of the shearing stress for the same case, and show at what point it is a maximum. Find where the shearing stress is zero. Discuss the case when $l_1 = l_2$.

14.—A GENERAL CASE.—A HORIZONTAL BEAM IS LOADED AT ANY NUMBER OF POINTS.

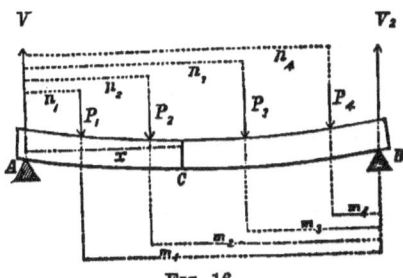

FIG. 16.

Let the notation be as in the figure.

Then $V = \dfrac{\Sigma Pm}{l}$ and $V_2 = \dfrac{\Sigma Pn}{l}.$

The general moment of stress is

$$Vx - P_1(x-n_1) - P_2(x-n_2), \&c. \dots \dots (38)$$

to include all the terms of P's in which x is less than the n with which it is joined in forming the moment.

The shearing stress is

$$S_t = V - P_1 - P_2 - P_3, -\&c. \dots \dots (39)$$

including all the loads between A and the section which is considered.

15.—OBLIQUE STRAINS.—BEAMS FIXED AT ONE END AND A STRESS APPLIED AT THE FREE END.

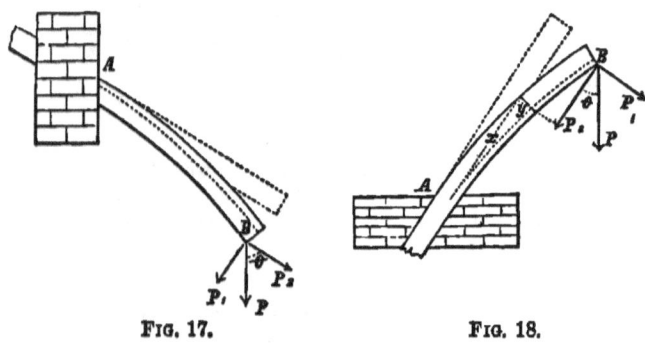

FIG. 17.   FIG. 18.

Let x be counted from the free end. Then the moment of transverse stress in both cases is

$$P_1 x.$$

The maximum moment is at the fixed end, and is

$$P_1 l.$$

In Fig. 18 the stress $P_2 = P \cos \theta$ tends directly to compress the piece, and in Fig. 17 P_2 tends to elongate it. If the beam is rectangular, and breadth b and depth d, the stress on a unit of section (square inch) in either case is $\dfrac{P_2}{bd}$, and this value must be deducted from the modulus of rupture in determining the strength of the piece. Hence we have for rectangular prismatic beams, at the dangerous section, in either case (see *Resistance of Materials*, Article 114, third edition):

$$P_,l = \tfrac{1}{6}\left(R - \frac{P_,}{bd}\right)bd^2,$$

or, $$P \sin \theta . l = \tfrac{1}{6}\left(R - \frac{P\cos\theta}{bd}\right)bd^2$$

EXAMPLES.—1. A prismatic beam whose length is 10 feet, is inclined at an angle of 30 degrees to the vertical, and is to sustain a weight of 1,000 pounds; the modulus of rupture (R) is 10,000 pounds, the breadth is 8 inches; required the depth for a coefficent of safety of $\tfrac{1}{10}$.

2. For the same length, weight, modulus of rupture, and coefficient of safety as in the preceding example; required the dimensions when the beam is square.

16.—INCLINED BEAM.—LET THE BEAM BE INCLINED TO THE HORIZONTAL, AND A WEIGHT, P, REST UPON IT AT ANY POINT.

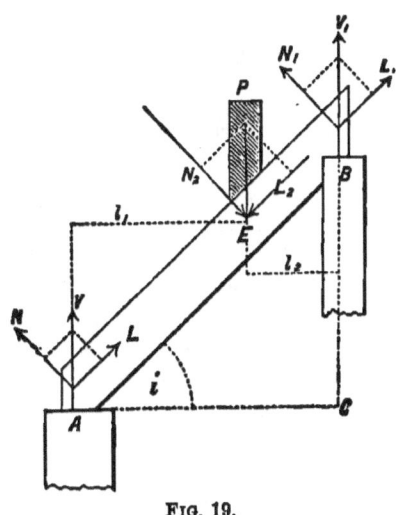

FIG. 19.

Let $l = AC =$ the horizontal distance between the centres of the bearings; and the other notation as in Figure 19.

We then have
$$V + V_1 = P, \text{ and}$$
by moments we have
$$Vl = Pl_2$$
$$\therefore V = \frac{l_2}{l} P; \text{ also}$$
$$V_1 = \frac{l_1}{l} P.$$

By resolving the applied force and reactions, we have
$$N = V \cos i = \frac{l_2}{l} P \cos i.$$
$$L = V \sin i = \frac{l_2}{l} P \sin i.$$
$$N_1 = V_1 \cos i = \frac{l_1}{l} P \cos i.$$
$$L_1 = V_1 \sin i = \frac{l_1}{l} P \sin i.$$
$$N_2 = P \cos i.$$
$$L_2 = P \sin i.$$

In this case there is no thrust on A tending to push it over, or similar pull on B; but the whole stress on A and B is vertical. But the strain due to the stress V_1 is oblique to the axis of the beam, and it may be resolved into two components, one of which (N_1) is normal to and the other (L_1) parallel to the axis of the beam. The same is true of P and V. The original strains are therefore oblique to the axis of the beam throughout the whole length. The reaction V_1 has a component, L_1, which *pulls* on the piece; and similarly L, one of the components of V, *pushes* on the piece. The sum of these efforts equals the longitudinal component L_2 of the weight P.

$$\therefore L_2 = L + L_1,$$
or, $$P \sin i = V \sin i + V_1 \sin i.$$
In a similar way we have
$$N_2 = N + N_1.$$

STRAINS ON SINGLE PIECES. 17

Each part of the beam or rafter $A\ E$ and $B\ E$, is now under substantially the same condition as those in the preceding problem. The dangerous section for prismatic beams will be at E, directly under the weight P. Hence we have for the moments of stress and strain, for rectangular sections at the dangerous section,

$$N \times AE = \tfrac{1}{6}\left(R - \tfrac{L}{bd}\right)bd^2;$$

or, $$\tfrac{l_1}{l} P \cos i \times \tfrac{l_1}{\cos i} = \tfrac{1}{6}\left(R - \tfrac{L}{bd}\right)bd^2$$

$$\therefore \tfrac{l_1 l_1}{l} P = \tfrac{1}{6}\left(R - \tfrac{L}{bd}\right)bd^2\ldots\ldots(40)$$

Similarly, we may find the strain by considering the reaction V_1, thus:

$$N_1 \times BE = \tfrac{1}{6}\left(R - \tfrac{L_1}{bd}\right)bd^2$$

$$\therefore \tfrac{l_1}{l} P \cos i \times \tfrac{l_1}{\cos i} = \tfrac{1}{6}\left(R - \tfrac{L_1}{bd}\right)bd^2$$

$$\therefore \tfrac{l_1 l_1}{l} P = \tfrac{1}{6}\left(R - \tfrac{L_1}{bd}\right)bd^2\ldots\ldots(41)$$

There is an apparent discrepancy in these results, the quantities within the parenthesis being different except when $L = L_1$. This may be explained by considering that if the weight P is divided in the ratio of V to V_1, and the two weights removed an infinitely short distance from each other, as in Fig. 20; then will the two longitudinal components of V_1 balance each other, being equal and opposite. The same is true of V. Hence, within the infinitely short space between a and b, there will be no longitudinal strain; and hence in this case we shall have for this section

$$\tfrac{l_1 l_1}{l} P = \tfrac{1}{6} R bd^2;$$

and which may be considered true for an *ideal* transverse section through the centre of action of P, Fig. 19. But this is an

2

ideal case. In Fig. 20, the strain immediately at the left of a is given by equation (40) and that at the right by equation (41). Hence in any practical case we should use equation (41), if L exceeds L_1; and equation (40) if the reverse be true.

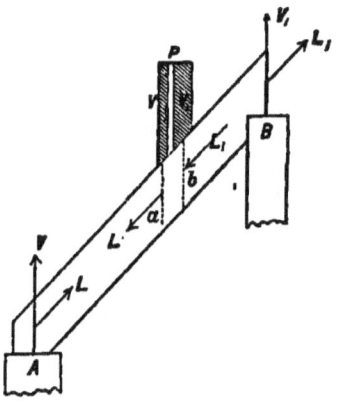

FIG. 20.

This may be still further illustrated by supposing that the beam is horizontal, and acted upon by the resolved forces of Fig. 19. Suppose in Fig. 21 that there is a pin at E to resist

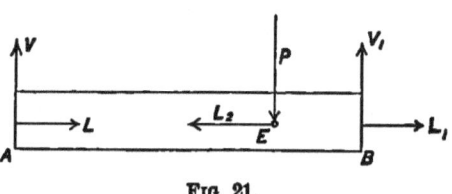

FIG. 21.

the strains which in Fig. 19 were induced by P. Then the force L_1 pulls along EB and finally against the pin E. Similarly, the force L pushes along AE, and finally is resisted by the pin. Hence, at the left-hand side of the pin, E, the pressure is $L_2 = L + L_1$; but just at the right of that point the pressure is L_1 as before stated.

17.—A LOADED BRACE.—LET A BEAM OR BRACE REST AGAINST AN ABUTMENT AT ITS LOWER END, AND AGAINST A VERTICAL WALL AT THE UPPER END, AND SUPPORT A WEIGHT, P, AT ANY POINT.

Figure 22 represents the case; for it makes no difference at the upper end whether it rests against a vertical wall, or post, or other vertical surface; or has a horizontal force applied there which is sufficient to keep it in the same position as a vertical surface.

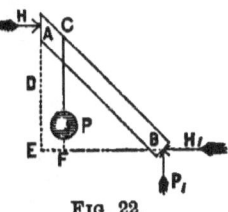

FIG. 22.

Let AB = the length of the brace (or rafter);
$D = AE$ = the rise;
$l = EB$ = the run, or horizontal projection;
P = the weight which is applied at C;
$nl = BF$ = the horizontal distance of the point of application of the weight P from B;
θ = the angle $BCF = BAE$;
P_1 = the reaction of the support B;
H_1 = the horizontal pressure at B;
$H_1 = H$ = the horizontal pressure at A.

It is a principle of statics that the sum of all the horizontal components of the applied forces, for equilibrium, is zero; and similarly for the vertical components.

Hence
$$P = P_1,$$
$$H = H_1,$$

Taking the moments about B, and we have
$$HD = Pnl$$
$$\therefore H = n P \frac{l}{D} = n P \tang \theta \ldots \ldots (42)$$

Taking the moments about A, and we have
$$P_1 l = H_1 D + (1-n) Pl,$$
in which make $P_1 = P$. Reduce, and we have
$$H_1 = n P \frac{l}{D} = n P \tang \theta \text{ as before.}$$

In a similar way, assuming that $H = H_1$, and we find $P = P_1$.

Resolving the force H into two others, one of which shall be perpendicular to the axis of the beam, and the other parallel to it, as shown in Fig. 23, and we have

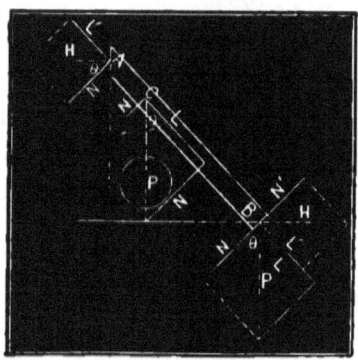

FIG. 23.

$$L' = H \sin\theta = n\,P\,\tan\theta \sin\theta = n\,P\,\frac{\sin^2\theta}{\cos\theta}.$$

$$N' = H \cos\theta = n\,P\,\tan\theta \cos\theta = n\,P \sin\theta$$

The components of P are

$$N = P \sin\theta \quad \dotfill (43)$$
$$L = P \cos\theta \quad \dotfill (44)$$

The compression due to the longitudinal components are:—

between A and C, $L' = n\,P\dfrac{\sin^2\theta}{\cos\theta}$; $\dotfill (45)$

between C and B,

$$L + L' = \frac{P}{\cos\theta}\Big[n \sin^2\theta + \cos^2\theta\Big] \dotfill (46)$$

The same result may be obtained by resolving the forces at the lower end.

18.—DISCUSSION OF THE PRECEDING CASE.

1. If the rafter is vertical $\theta = 0$, and equation (45) gives $L = 0$, and equation (46) gives $L = P$; and (42) gives $H = 0$.

2. If the weight is placed at A, we have $n = 1$, and
equation (42) gives $H = P \tan\theta$, and
equation (46) gives $L + L' = P \sec\theta$.

3. If P is placed at the lower end, or at B, we have $n = 0$, and equations (42) and (45) give zero, while (46) gives $P \cos \theta$. This does not give a strain, but is simply the longitudinal component of P.

4. If P is applied at the middle, $n = \frac{1}{2}$, and we have

$$H = \tfrac{1}{2} P \tang \theta \dotfill (47)$$
$$L' = \tfrac{1}{2} P \tang \theta \sin \theta \dotfill (48)$$
$$L + L' = \frac{P}{\cos \theta} \left[\tfrac{1}{2} \sin^2 \theta + \cos^2 \theta \right] \dotfill (49)$$

5. If the rafter is horizontal the several strains become infinite.

19.—The moment of transverse stress between A and C is
$$N' x' = n P x' \sin \theta$$
in which x' is reckoned from A. And between C and B it is
$$(N - N') x = (1 - n) P x \sin \theta$$
in which x is reckoned from B.

Hence, for rectangular beams, we have for rupture between A and C,
$$n P x' \sin \theta = \tfrac{1}{6} \left(R - \frac{n P \sin^2 \theta}{bd \cos \theta} \right) bd^2$$

The dangerous section is at C, for which we have
$$n P \times AC \sin \theta = \tfrac{1}{6} \left(R - \frac{P (\cos^2 \theta + n \sin^2 \theta)}{bd \cos \theta} \right) bd^2 \dots (50)$$

PROBLEMS.—1. A beam whose length is 20 feet, breadth 6 inches, depth 12 inches, is inclined at an angle of 60° to the vertical; required the weight which it will safely sustain when placed at the centre. Call R 1,000 pounds.

2. Required the load at the centre which the same beam will sustain if horizontal.

3. Required the load which a beam of the same breadth and depth as in example 1, and whose length is the horizontal projection of that beam, will sustain if placed at the centre.

20.—LOADED RAFTER.—CASE OF A RAFTER WHICH IS LOADED UNIFORMLY OVER ITS WHOLE LENGTH.

Let $w =$ the load per unit of length,
$x =$ any distance measured from A,
$W =$ the total load on the rafter, and
$l =$ the horizontal projection of the rafter.

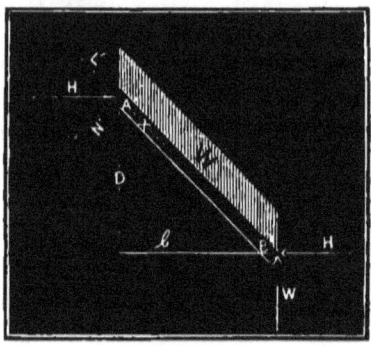

Fig. 24.

Taking the moments about B we have

$$HD = W \cdot \tfrac{1}{2} l;$$
$$\therefore H = \tfrac{1}{2} W \frac{l}{D} = \tfrac{1}{2} W \tan \theta \ldots \ldots \ldots (51)$$

which according to Eq. (47) is the same as if the whole load were concentrated at the centre.

The load on x is wx and according to equation (44) the longitudinal component of it is

$$L = wx \cos \theta, \ldots \ldots \ldots \ldots \ldots (52)$$

and the longitudinal component of the thrust at the upper end is

$$L' = H \sin \theta = \tfrac{1}{2} W \tan \theta \sin \theta ; \ldots \ldots (53)$$

which according to equation (48) is the same as if the whole load were concentrated at the centre.

Hence the total compression at any section whose distance from A is x, is

$$L + L' = wx \cos \theta + \tfrac{1}{2} W \tan \theta \sin \theta \ldots \ldots (54)$$

At the lower end x becomes AB and we have

$$L + L' = \frac{W}{2 \cos \theta} \left[2 \cos^2 \theta + \sin^2 \theta \right] = \frac{W}{2 \cos \theta} \left[1 + \cos^2 \theta \right] \ldots (55)$$

At the middle $x = \tfrac{1}{2} AB$ and the longitudinal compression at the middle is

$$\tfrac{1}{2} W \sec \theta \ldots \ldots \ldots \ldots (56)$$

The normal component of H is
$$N' = H \cos \theta = \tfrac{1}{2} W \tan \theta \cos \theta = \tfrac{1}{2} W \sin \theta.$$
The normal component of the load wx is
$$N = wx \sin \theta.$$
Hence we have for the equation for rupture at any point, for rectangular sections,
$$N'x - \tfrac{1}{2} Nx = \tfrac{1}{6}\left[R - \frac{L+L'}{bd}\right]bd^2$$
or
$$Wx - wx^2 = \frac{1}{3 \sin \theta}\left[R - \frac{2wx \cos^2 \theta + W \sin^2 \theta}{2\, bd \cos \theta}\right]bd^2 \quad (57)$$
From equation (57) we have
$$b = \frac{3 \sin \theta}{Rd^2}\left[Wx - wx^2\right] + \frac{wx \cos \theta}{Rd} + \tfrac{1}{2}\frac{W \sin^2 \theta}{Rd \cos \theta}.$$
If the depth be constant we may easily find from this equation the position of the dangerous section. By differentiating and placing equal to zero we have
$$\frac{db}{dx} = \frac{3 \sin \theta}{Rd^2}\left[W - 2wx\right] + \frac{w \cos \theta}{Rd} = 0$$
$$\therefore x = \tfrac{1}{2} AB + \tfrac{1}{6} d \cot \theta.$$
This value in Eq. (57) enables us to find the proper value for b.

It is much more difficult to find the position of the dangerous section when b is constant and d variable.

EXAMPLES.—1. A rafter whose length, AB, is 20 feet, breadth 2 inches, depth 6 inches, inclination to the vertical 60 degrees, is loaded uniformly over its whole length. If $R = 1,000$ pounds, required the load which it will sustain.

2. A rafter whose length is 20 feet, breadth 2 inches, inclination to the horizontal 25 degrees, $R = 800$ pounds, is required to carry 40 pounds per foot of its length; required its depth.

21.—STRENGTH OF PILLARS.—The general formula for determining the strength of long prismatic pieces which are subjected to longitudinal compression is
$$P = \frac{\pi^2}{l^2} EI. \quad\quad\quad\quad (58)$$

The strength of square pillars which rest on a flat base and the upper end of which is also flat and secured in its position, may be determined from the following theoretical formula.

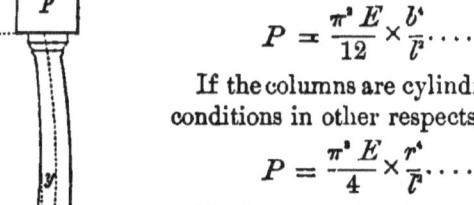

$$P = \frac{\pi^2 E}{12} \times \frac{b^4}{l^2} \dots \dots \dots (59)$$

If the columns are cylindrical, with the same conditions in other respects, we have

$$P = \frac{\pi^2 E}{4} \times \frac{r^4}{l^2} \dots \dots \dots (60)$$

For hollow cylindrical columns, we have

$$P = \tfrac{1}{4} \pi^2 E \frac{r^4 - r_1^4}{l^2} \dots \dots \dots (61)$$

Fig. 25.

These formulas, according to Navier and Weisbach, should be used only when the length exceeds 20 times the diameter for cylindrical columns, or 20 times the least thickness for rectangular columns; and Navier says that only $\tfrac{1}{10}$ the calculated weight should be used for safety for wooden columns, and $\tfrac{1}{4}$ to $\tfrac{1}{5}$ for iron ones, but Weisbach uses $\tfrac{1}{10}$ in the case of cast iron and of wood.

Practical men are disposed to distrust these formulas as being too theoretical. They use instead thereof Hodgkinson's empirical formulas, which were deduced from a long series of experiments; or Gordon's formula, which was deduced from the same experiments. The results given by these formulas do not differ very largely from those given above.

22.—GORDON'S RULES FOR THE BREAKING WEIGHT IN POUNDS, OF IRON PILLARS WITH FLAT ENDS, THE ENDS BEING SECURED SO THAT THEY CANNOT MOVE LATERALLY, AND EVENLY DISTRIBUTED OVER THE END SURFACES.

The thickness of the metal in hollow columns not to exceed $\tfrac{1}{6}$ to $\tfrac{1}{8}$ of the external diameter.

HOLLOW CYLINDRICAL PILLARS.

$$\text{FOR CAST IRON, } P = \frac{80{,}000\, A}{1 + \dfrac{l^2}{800\, D^2}} \dots \dots \dots (62)$$

STRAINS ON SINGLE PIECES.

in which A = the sectional area of the metal in square inches,
l = the length of the pillar in inches,
D = the outer diameter in inches,
P = the breaking load in pounds, and
80,000 = the crushing strength of cast iron, which is a low value, and hence safe.

FOR WROUGHT IRON,

$$P = \frac{36,000\,A}{1 + \dfrac{l^2}{3,000\,D^2}} \quad \dots\dots\dots\dots\dots\dots\dots(63)$$

SQUARE HOLLOW PILLARS.

FOR CAST IRON, $\quad P = \dfrac{80,000\,A}{1 + \dfrac{l^2}{533\,b^2}}$

FOR WROUGHT IRON, $\quad P = \dfrac{36,000\,A}{1 + \dfrac{l^2}{6,000\,b^2}}$

SOLID PILLARS.

FOR CAST IRON, $\quad P = \dfrac{80,000\,A}{1 + \dfrac{l^2}{266\,b^2}} \quad \dots\dots\dots\dots\dots(64)$

FOR WROUGHT IRON, $\quad P = \dfrac{36,000\,A\,{*}}{1 + \dfrac{l^2}{3,000\,b^2}} \quad \dots\dots\dots\dots(65)$

23.—MR. C. SHALER SMITH'S FORMULA for the strength of white or yellow pine square pillars.

$$P = \frac{5,000}{1 + \dfrac{0.004\,l^2}{b^2}} \quad \dots\dots\dots\dots\dots(66)$$

---

* For tables computed from these formulas, see *Trautwine's Engineer's Pocket-Book*, pp. 238, 239 and 240.

24.—HODGKINSON'S FORMULAS.

TABLE
For the absolute strength of columns.

In which P = crushing weight in gross tons,
  d = the external diameter, or side of the column in inches,
  d_1 = the internal diameter of the hollow in inches, and
  l = the length in feet.

Kind of Column.	Both ends rounded, the length of the column exceeding fifteen times its diameter.	Both ends flat, the length of the column exceeding thirty times its diameter.
	TONS.	TONS.
Solid Cylindrical Columns of cast iron	$P = 14.9 \dfrac{d^{3.76}}{l^{1.7}}$	$P = 44.16 \dfrac{d^{3.55}}{l^{1.7}}$
Hollow Cylindrical Columns of cast iron	$P = 13 \dfrac{d^{3.76} - d_1^{3.76}}{l^{1.7}}$	$P = 44.34 \dfrac{d^{3.55} - d_1^{3.55}}{l^{1.7}}$
Solid Cylindrical Columns of wrought iron	$P = 42 \dfrac{d^{3.76}}{l^2}$	$P = 133.75 \dfrac{d^{3.55}}{l^2}$
Solid Square Pillar of Dantzic oak	....	$P = 10.95 \dfrac{d^4}{l^2}$
Solid Square Pillar of red dry deal	....	$P = 7.81 \dfrac{d^4}{l^2}$

The above formulas apply only in cases where the length is so great that the column breaks by bending and not by simple crushing. One-fifth of the results given by these formulas, is considered safe in practice.

If the length of the column is less than that given in the table, and more than four or five times its diameter, the strength is found by the following formula:

$$W = \frac{P \cdot CK}{P + \frac{3}{4}CK} \quad \ldots \ldots \ldots \ldots \ldots (67)$$

in which P = the value given in the preceding table,
  K = the transverse section of the column in square inches,
  C = the modulus for crushing in tons (gross) per square inch, and
  W = the strength of the column in tons (gross).*

---

* James B. Francis, C.E., has published a set of tables which gives the strength of cast-iron columns, of given dimensions, by means of equation (67), and also by those given in the above table.

STRAINS ON SINGLE PIECES. 27

25.—WEIGHT OF PILLARS.

TABLE

Of the weights in pounds of pillars in terms of their lengths (l) in feet, when loaded to one-fifth their crushing strength (P) in pounds.

Kind of Pillar.	Weight in pounds.	
	Both ends rounded. $l > 15\, d.$	Both ends flat. $l > 30\, l.$
Solid Cylindrical Column of cast iron.	$0.023926469\ (P.l^{2.55})^{\frac{1}{1.55}}$	$0.009321706\ (P.l^{2.475})^{\frac{1}{1.475}}$
Hollow Cylindrical Columns of cast iron, $d_1 = nd.$	$0.024392078\dfrac{1-n^2}{(1-n^{3.75})^{\frac{1}{1.55}}} \times (P.l^{2.55})^{\frac{1}{1.55}}$	$0.009300164\dfrac{1-n^2}{(1-n^{3.55})^{\frac{1}{1.475}}} \times (P.l^{2.475})^{\frac{1}{1.475}}$
if $n = 0.98$	$0.003881655\ (P.l^{2.55})^{\frac{1}{1.55}}$	$0.001658133\ (P.l^{2.475})^{\frac{1}{1.475}}$
if $n = 0.95$	$0.006001775\ (P.l^{2.55})^{\frac{1}{1.55}}$	$0.002489827\ (P.l^{2.475})^{\frac{1}{1.475}}$
if $n = 0.925$	$0.007265678\ (P.l^{2.55})^{\frac{1}{1.55}}$	$0.002987882\ (P.l^{2.475})^{\frac{1}{1.475}}$
if $n = 0.90$	$0.008396144\ (P.l^{2.55})^{\frac{1}{1.55}}$	$0.003406063\ (P.l^{2.475})^{\frac{1}{1.475}}$
if $n = 0.875$	$0.009373430\ (P.l^{2.55})^{\frac{1}{1.55}}$	$0.003773531\ (P.l^{2.475})^{\frac{1}{1.475}}$
if $n = 0.85$	$0.010261367\ (P.l^{2.55})^{\frac{1}{1.55}}$	$0.004106903\ (P.l^{2.475})^{\frac{1}{1.475}}$
if $n = 0.80$	$0.011862713\ (P.l^{2.55})^{\frac{1}{1.55}}$	$0.004702651\ (P.l^{2.475})^{\frac{1}{1.475}}$
$n = 0.75$	$0.013297905\ (P.l^{2.55})^{\frac{1}{1.55}}$	$0.005233352\ (P.l^{2.475})^{\frac{1}{1.475}}$
Solid Cylindrical Columns of Wrought Iron,	$0.014115831\ (P.l^{2.55})^{\frac{1}{1.55}}$	$0.004993604\ (P.l^{2.475})^{\frac{1}{1.475}}$
Square Column of Dantzic Oak.	(Cubic foot weighs 47.24 pounds.)	$0.001223770\ (P^{\frac{1}{2}}.l)$

Part 2.

TRUSSED BRIDGES.

CHAPTER I.

DEFINITIONS AND DATA

26.—DEFINITIONS.—A STRUCTURE is an assemblage of pieces so joined that the whole may act in sustaining weights, or the pressure of water, or the pressure of wind, or other physical pressures. The points at which the pieces are joined are usually called **JOINTS**. The pieces between the joints are usually rigid, although this is not necessarily the case, for a flexible piece as a rope, wire, or chain, will resist a *pull* as effectually as a bar of iron. Structures may be composed of earth, or masonry, as well as of iron, wood, or other materials.

A FRAME is a *structure* which is usually composed of wood, or iron, or both combined, the pieces being joined at points, and the whole so constructed, arranged and secured that it may carry loads, or resist the action of external forces.

A BEAM is a simple or compound piece which is usually supported at its ends for sustaining a transverse or oblique stress. Beams are sometimes built up, in which case they may be called compound beams.

A GIRDER is either a simple beam, or a framed assemblage of pieces so constructed that it may carry loads when it is supported at its ends or at other intermediate points.

A TRUSS is a framed *girder*. It is a *frame*, but a *frame* implies more than a *truss*. A *frame* is a more general term. We speak of a *framed* house, but not of a *trussed* house.

Framing relates to the joining of the parts, *trussing* to arranging the parts so that the frame may not be distorted under the action of the forces to which it is to be subjected. Roofs are often trussed. Many bridges are called trussed bridges. Trussing generally implies *bracing*. Thus, a trussed partition is a braced partition. In bridges and in roofs, *ties* often take the place of *braces*. The distinction between *girders* and *trusses* is not always made, although the tendency seems to be to use the former as a dignified name for a beam, and the latter as applicable to braced and tied structures.

A **SEMI-TRUSS** is a truss which has no support at one end, as in Fig. 56.

A **BRIDGE** is a *structure* over a river or other body of water for supporting moving loads. It usually connects two roads in such a way as to form a continuous road. A *bridge* may be a *frame*, or it may be an arch, or it may be composed simply of beams, or it may be suspended by cables. These several forms give rise to several classes of bridges. For the purposes of analysis it is convenient to divide them into—

1. TRUSSED BRIDGES.
2. TUBULAR BRIDGES.
3. SUSPENSION BRIDGES.
4. ARCHED BRIDGES.
5. COMPOUND STRUCTURES.

Each of these admit of several subdivisions, as will hereafter appear.

A **VIADUCT** is a *structure* for carrying a roadway over low ground, where there is not necessarily any water. The structure may be in all respects like a bridge.

A **CHORD** is the upper or lower member in a *truss*. It extends from end to end of the structure. There are usually two chords, an upper and a lower chord. These may be parallel, as in Figs. 69 and 88, or the upper one may be curved (arched) and the lower one horizontal, or both may be curved. These pieces by some English writers are called *booms;* and by others, *stringers*. The lower chord is often called a *tie*. The upper chord is sometimes called a *straining* beam.

A **TIE** is a piece which connects two parts and is subjected to tension.

A **STRUT** is a general term which is applied to a piece in a truss which is subjected to compression. In proportioning it, it is treated as a *pillar*. In its more restricted sense, it is a *short* piece which is subjected to compression.

A **TIE-STRUT** or **STRUT-TIE** is a piece which may be subjected to tension and compression at different times, for different conditions of loading.

A **BRACE** is an inclined piece which is subjected to compression. It is an inclined *strut*. In bridges, braces are sometimes distinguished as *main-braces* and *counter-braces*. This distinction is quite unnecessary in an analytic point of view, as will be seen hereafter, but it is so common in practice that it will not do to ignore it.

A **MAIN-BRACE** is a brace which inclines from the end of a truss towards the centre, as in Fig. 26.

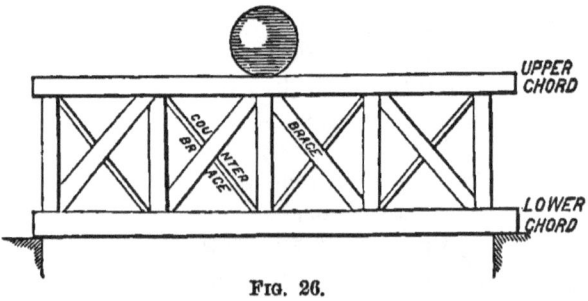

FIG. 26.

A **COUNTER-BRACE** is one which inclines from the centre and toward the ends. In the same panel the counter-brace inclines the opposite way from the main-brace. See Fig. 26.

A **TIE-BRACE**, performs the office of both main and counter-brace. It is the same as a **TIE-STRUT**.

A **KING-POST** is a term applied to a truss in which there is a central tie (or post) and two braces resting against it as in Fig. 28. This is an old and familiar term, and came into use when the central piece was made of wood and resembled a post, although its office was that of a tie. A *braced post* or a *braced tie*, would seem to be a more appropriate mechanical term.

A **QUEEN-POST TRUSS** is a truss in which there are two posts (or ties) and against which rest two braces. The horizon tal piece between the upper ends of the posts is often called a *straining beam*. See Fig. 49. This is also an old term and has become familiar with long usage, but, mechanically speaking, it would be much more appropriate to call it a **TRAPEZOIDAL TRUSS**. There are other forms of trusses which in outline resemble a trapezoid, and it does not seem necessary to separate this case from the more general one. It may, however, be desirable to do so on account of its frequent use in roofs, and bridges of short span.

27.—THE LOAD. The load on a bridge consists of two parts; the first part being permanent and due to the weight of the bridge, is sometimes called the "dead load;" and the second part being "temporary," is composed of moving trains of cars, or teams, or persons, or other loads. The latter is sometimes called "the moving load," or "the live load," or "the surcharge." The **LIVE LOAD** may be assumed, or is given in the original data, but the **PERMANENT LOAD** must be assumed, and can be accurately determined only by successive approximations. For it is dependent upon the thing which we seek viz., the dimensions of the parts of the bridge. We therefore, at first, assume such a value for the total weight as is indicated by other similar structures, and after the dimensions of all the parts have been computed, the weight is calculated from these dimensions; and if the assumed weight does not largely exceed the computed weight, it will not be necessary to review the calculation.

I have examined the weights of several wooden railroad bridges, and I find that for spans of about 150 feet, they average about 1,200 pounds per linear foot. We may *assume* for iron bridges of spans from 100 to 180 feet, from a half to three-fourths of a ton, and from 180 to 250 feet, from three-fourths to one ton per foot; and for spans of 300 feet, about 1¼ ton net.

The Conway tubular bridge, England, weighs about 3,000,000 pounds and is 400 feet long, and hence weighs about 7,500 pounds per linear foot.

The **LIVE LOAD** is generally assumed to equal or slightly exceed the greatest load which we think will ever be placed upon the bridge. For a railroad bridge the heaviest load would be that of a train of locomotives, extending from end to end of the bridge—or at least over one span of it. If the bridge is long it is not probable that so heavy a load would ever be placed upon it, but it makes the computation much more simple to *assume* this load, than it does to consider one or two locomotives followed by a train of cars which is much lighter; and the error, if any, which results from the *assumption*, is on the safe side.

English and American engineers assume that a train of locomotives will weigh one ton (gross) per foot of length. Morin says that the French locomotives weigh about $\frac{4}{5}$ of a ton per linear foot.* The maximum weight of a train of merchandise on the road from Paris to Lyons is given as 1674 kilog. per metre of length,* or more than one-half ton (net) per foot.

In *Traité Théoretique et Pratique* on the construction of metallic bridges by MM. L. Motinos and C. Ponnier, p. 60, the following values are given; the length in metres, the weights in French tons.

Length of the Span.	4 m	6 m	10 m	15	20	25	30	40½	60.
Weight per metre of length uniformly distributed. tons	8	7	6	5	4.7	4.5	4.5	4.5	4

Which in English units gives, to the nearest tenth,

Length of Span in feet.	13 ft.	19½	32½	48½	65	81½	97½	130	195
Weight per foot of length uniformly distributed. tons	2.5	2.2	1.8	1.5	1.4	1.3	1.3	1.3	1.1

So that for long bridges they recommend rather more than one ton (net) per foot. It is advisable in all cases to allow a larger coefficient for safety for short spans than for long ones, for they are liable to be subjected to much greater concentrated loads and to more severe local shocks. As much as 16,000 or 18,000 pounds may rest upon the driving wheels

---

* Morin, *Résistance des Matériaux*, pp. 333 and 334.

SAFE LOAD. 33

of an engine, which load extends over a few feet only. In this country one ton per foot is commonly used.

For common road bridges we may safely *assume* that the maximum load is what it would be if completely covered with men.

For suspension bridges the French engineers use forty pounds per square foot.* Navier recommends 42 pounds per square foot.† But a crowd of persons will weigh much more than this. Picked men closely packed in one experiment weighed eighty-four pounds per square foot. An experiment by Mr. Nash, architect of Buckingham Palace, in which he *wedged* men together as closely as possible, secured 120 pounds per square foot.‡

28.—SAFE LOAD. A structure should not be so heavily loaded as to damage the elasticity of the materials which compose it. It is impossible to tell the exact load which the structure can sustain without passing this limit, but considerations of safety and durability demand that one should keep on the safe side. Hence it is often the case in practice that bridges are made 1½ to 2 times as strong as would be absolutely necessary if the materials were all of a known standard quality and the workmanship practically perfect. In order to make the structure safe against all these contingencies, it is made several times as strong as is necessary for sustaining the load at the crushing limit.

29.—THE FACTOR OF SAFETY is the ratio of the computed strain to the actual strain; or in other words it is the ratio of the load which would just crush the structure to the assumed load. There is no absolute rule for determining the proper value of this *factor*. Its value is assumed *arbitrarily* by the engineer, although its extreme limits may be determined by experiment and observation. For instance, experiment shows that the *factor* for the tension of wrought iron should never be so small as 2; and when the load is applied without shock it ought not to be less than 3. It seems evident also that it is unnecessary to

---

* Mahan's *Civ. Eng.*, p. 257.
† Weisbach's *Mech.* and *Eng.* (3d Ed.), Vol. II. p. 64.
‡ Trautwine's *Eng. Pocket-Book*, p. 297.

allow so large a *factor* as 10, or 8, or even 6 for wrought iron where there is no shock. But in all practical cases in bridge construction there is some shock due to passing loads, and hence we may assume a *factor* as large as 6, as is shown in the following list. Between these limits its value depends chiefly upon the *choice* of the engineer. From 4 to 5 is a very common value for wrought iron subjected to tension or cross-strain, from 4 to 6 for cast iron, and 10 for wood.

The following are the values used by several engineers and authors:—

Factors.
Messrs. May and Grissel..................................3
Mr. Brunell...3 to 5
Messrs. Rasbrick, Benton and others.....................6
Mr. Hawkshaw...7
Mr. Glyn..10
Bow for wrought-iron beams...........................3.5
Weisbach, for wrought iron.........................3 to 4
Vicat for wire suspension bridges.............more than 4
Fink, iron truss bridges { for posts and braces........5 to 6
                         { for cast-iron chords............7
Fairbairn, for cast-iron beams.....................5 to 6
C. Shaler Smith, compression of cast iron.................5
Rankine and others, for cast-iron beams...............4 to 6
Mr. Clark in Quincy Bridge, lower chord ....6 to 7
Washington A. Roebling, for suspension cables............6
The Detroit Bridge and Iron Company................5 to 6
Morin, Vicat, Weisbach, Rondelet, Navier, Barlow, and many others say that for a wooden frame it should not be less than...10
For stone, for compression......................10 to 15

Mr. Fairbairn deduced the following conclusions from experiments which were made by him in regard to beams and girders, whether plain or tubular. "The weight of the girder and its platform should not in any case exceed one fourth the breaking weight, and that only one-sixth of the remaining three-fourths of the strength should be used by the moving load." According to this statement the maximum

load, including the live and dead loads, may equal, but should not exceed,
$$\tfrac{1}{4} + \tfrac{1}{8} \text{ of } \tfrac{3}{4} = \tfrac{3}{8}$$
of the breaking load. Hence the *factor of safety* must not be less than 2.66 when the above conditions are fulfilled. This value is, however, evidently smaller than is thought advisable by most engineers.

The rule adopted by the Board of Trade, England, for railroad bridges is " to estimate the strain produced by the greatest weight which can possibly come upon a bridge throughout every part of the structure, which should not exceed *one-fifth the ultimate strength of the metal.*" They also observed that ordinary road bridges should be proportionately stronger than ordinary railroad bridges.

30.—ABSOLUTE MODULUS OF SAFETY.—Sometimes an arbitrary value is assumed for the maximum strain in pounds per square inch to which the material may be subjected. Such a value is called the *absolute modulus of safety.*

The following values are generally assumed for the *modulus of safety.*

	Pounds per square inch.
Wrought iron, *for tension or compression*, from	10,000 to 12,000
Cast iron, for *tension*, from	3,000 to 4,000
Cast iron, for *compression*, from	15,000 to 20,000
Wood, *tension or compression*, from	850 to 1,200
Stone, *compression*, granite, from	400 to 1,200
quartz, from	1,200 to 2,000
sandstone, from	300 to 600
limestone, from	800 to 1,200

The practice of French engineers,* in the construction of bridges, is to allow 3.8 tons (gross) per inch upon the cross-section, both for tension and compression of wrought iron.

The Commissioners on Railroad Structures, England, established the rule that the maximum tensile strain upon any part of a wrought-iron bridge should not exceed five tons (gross) per square inch.

31.—EXAMPLES OF STRAINS THAT HAVE BEEN USED IN PRACTICAL CASES. The margin of safety that has been used

---
* Am. R. R. Times, 1871, p. 6.

in various structures may or may not serve as guides in designing new structures. If the margin for safety is so small that the structure appears to be insecure and gives indications of failure, it evidently should not be followed. It serves as a warning rather than as a guide. If the margin is evidently excessively large, demanding several times the amount of material that is necessary for stability and durability, it is not a guide. Any engineer or mechanic, without regard to scientific skill or economy in the use of materials, may err in this direction to any extent. But if the margin appears reasonably safe, and the structure has remained stable for a long time, it serves as a valuable guide, and one which may safely be followed under similar circumstances. Structures of this kind are practical cases of the approximate values of the inferior limits of the *factors of safety*. The following are some practical examples :—

IRON TRUSSED BRIDGES.

NAME OF THE BRIDGE.	TENSION. Tons per square inch.	COMPRESSION. Tons per square inch.
Passaic (*Lattice*)	5¼ to 6	4¼ to 5¼
Place de l'Europe (*Lattice*)	4	3¼
Canastota (*N. Y. C. R. R.*) (*Lattice*)	5	4
Newark Dyke (*Warren Girder*)	5	5
Boyne Viaduct (*Lattice*)	5	
Charing Cross (*Lattice*)	5	4
	Pounds per square inch.	Pounds per square inch.
St. Charles, Mo. (*Whipple Truss*)	12,000	12,000
Louisville, Ky. (*Fink Truss*)	7,000 to 12,000	½ to ⅔ the strength
Keokuk and Hannibal	9,251	8,962
Quincy Bridge	10,000	Factor of safety, 5
Kansas City Bridge		
Hannibal Bridge (*Quadrangular Truss*)	Factor of safety, 5	Factor of safety, 5

WOODEN BRIDGES.

NAME OF THE BRIDGE.	MAXIMUM STRAIN.
Cumberland Valley R. R. Bridge	635 pounds per square inch.
Portage Bridge (*N. Y. & E. R. R.*)	Factor of safety, 20.

EXAMPLES OF BRIDGES. 37

SUSPENSION BRIDGES.

NAME OF THE BRIDGE.	Span in feet.	Strain in tons per square inch. From Bridge.	Strain in tons per square inch. Bridge and Load.	Factor of safety.
Menai	580	4.21	8.00	3.9
Hammersmith	422¼	5.38	9.36	3.3
Pesth	666	5.01	8.11	3.9
Chelsea	384	4.36	8.07	3.9
Clifton	702½	2.90	5.03	6.4
Niagara	821	6.70	8.40	5.3
Suspension Aqueduct, Pittsburgh, Pa. 7 spans each,	160	....	....	4.0
Cincinnati Bridge	1,057	9.1	11.7	6.2
East River	1,600	....	....'	6.0
Highland (*proposed*)	1,600	....	....	6.0

TUBULAR BRIDGES.

NAME OF BRIDGE.	SPAN. Feet.	FOR WEIGHT OF BRIDGE AND LOAD.	
		Tension. Tons.	Compression. Tons.
Conway	400	6.85	5.03
Britannia (Central span)	460	3.00	....
Penrith (Tubular Girder)	...	4.75	4.25

CAST IRON ARCHES.

NAME OF THE ARCH.	SPAN.		VERSED SINE.		STRAIN PER SQUARE INCH IN TONS.
	feet.	inches.	feet.	inches.	
Austerlitz	106	0	10	7	2.78
Carrousel	152	2	16	1	1.46
St. Denis	102	5	11	4	1.37
Nevers	137	9	15	0	1.90
Rhone	197	10	16	5	2.37
Westminster	120	0	20	0	3.00

STONE ARCHES.

NAME OF THE ARCH.	Span in feet.	Versed sine in feet.	Pressure per square inch in pounds at the key.	Factor of safety at the point of greatest strain.
Wellington	100	15	175	11.3
Waterloo (9 *Arches*)	120	35	151	20.0
Neuilly	128	32	172	11.6
Taaf (*South Wales*)	140	25	244	8.0
Turin	147	18	203	10.2
London	152	38	215	14.0
Chester	200	42	349	8.6

CAST STEEL ARCH.

NAME OF ARCH.	SPAN, feet.	FACTOR OF SAFETY.
Illinois and St. Louis Bridge	515	6+

STONE FOUNDATIONS.

	FACTOR OF SAFETY.
Pillars of the Dome of St. Peter's (*Rome*)	16
" " " St. Paul's (*London*)	14
" " " St. Geneviève (*Paris*)	7.6
Pillars of the Church Toussaint (*Angers*)	10
Merchants' Shot Tower (*Baltimore*)	4.8
Lower courses of Britannia Bridge	31
Lower courses of the piers of Neuilly Bridge (*Paris*)	15.8
Foundation of St. Charles' Bridge (*Missouri*)	12 to 14
Foundations of East River Bridge *	10 to 20

* "In the stonework the pressures vary from 8 to 26 tons per square foot. Stone used is granite, selected samples of which have borne a crushing strain of 600 tons per square foot. Some will not bear over 100 tons per square foot. The general average is necessarily much less than that of the best specimens."—*Statement of the Chief Engineer*, Washington A. Roebling.

32.—PROOF LOAD.—The proof load is a trial load. It is intended as a practical test of a theoretical structure.

It generally exceeds the greatest load that it is ever intended to put upon the structure, but it should not be so great as to impair its elasticity. If the *proof load* is much in excess of the load which will ordinarily be placed upon the structure, it should remain on but a short time, and should be put on and removed in such a way as to avoid shocks as much as possible. *Excessively* severe *proof* strains may do much harm by permanently damaging the resisting properties of the materials.

33.—FRAMING. The art of *framing* pertains chiefly to the manner of joining the parts of solid materials so as to resist strains. This is a very important item especially in wooden structures. I have seen frames in which the pieces were so imperfectly secured to each other at their ends that they would fail there long before they were strained to the amount which they were expected to carry, making a frame very weak which in all other respects was very strong. Some of the approved methods for joining the ends will be indicated hereafter. In making the analysis of structures we shall assume that they are properly joined, and that they will yield only by their elasticity for such loads as they are intended to carry.

It is worthy of note, since joints cannot be made perfect, that where several joints are involved in carrying a strain, if one is over-strained, it will yield by its elasticity and thus bring into action others which were less strained. Were it not for this principle it would be exceedingly difficult to make large *compound* structures which would be durable.

THE SIMPLE GIRDER.

34.—THE SIMPLE GIRDER may be composed of a single piece of wood, iron, steel, or other metal, or it may be composed of several pieces firmly secured together so as to act like a single piece. Large beams and arches have been made of planks or even boards firmly spiked together. Some prefer built beams of this kind to solid ones, because they can select their timber and be certain that it is all sound, whereas it is difficult to secure large solid pieces, as some parts are liable to

TREATISE ON BRIDGES.

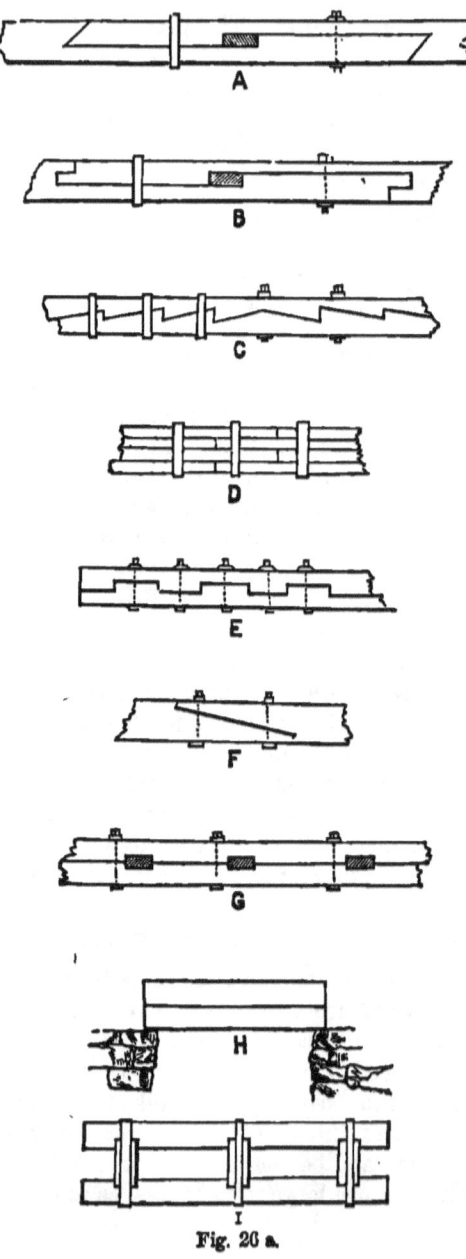

Fig. 26 a.

be *shaky* while the greater part is sound, or they may have internal defects. A built beam cannot be stronger than a solid one (as some have supposed), when the material in the solid one is as good throughout as that in the built one. If the pieces which compose the built one are large, they must be framed together. Figure (26 a) shows some of the approved modes of joining or splicing timbers.

If two equal timbers are simply placed over each other, as in H, Fig. (26 a), the strength of the two is double that of one, but if they can be firmly joined together so as to act like a single beam of the same depth as the two, the strength will be *four* times that of a single one.

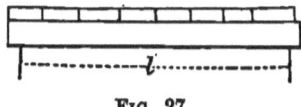

FIG. 27.

If a **BRIDGE** be made of several beams placed side by side as in Fig. 27, we may determine the breadth or depth of each as follows:—

Let $W =$ the total uniform load;
$l =$ the span;
$R =$ the modulus of rupture;
$b =$ the breadth of each beam;
$d =$ the depth of each beam; and
$n =$ the number of beams.

Then according to equation (19) we have

$$\tfrac{1}{8} Wl = \tfrac{1}{6} n \, Rbd^2$$

$$\therefore b = \tfrac{3}{4} \frac{Wl}{nRd^2}; \text{ or} \dots\dots\dots\dots\dots\dots (67\ a)$$

$$d = \tfrac{1}{2} \sqrt{\frac{3\,Wl}{nRb}} \dots\dots\dots\dots\dots\dots\dots\dots (67\ b)$$

If the depth of each beam is r times the breadth we have

$$\tfrac{1}{8} Wl = \tfrac{1}{6} nr^2 Rb^3$$

$$\therefore b = \sqrt[3]{\tfrac{3}{4} \frac{Wl}{nr^2 R}}$$

In the same way the dimensions of floor joists may be determined.

EXAMPLES. 1. The load on a floor is fifty pounds per square foot. The joists are twenty feet long, twelve inches deep, and are sixteen inches from centre to centre. What must be their breadth if $R = 800$ pounds?

2. The joists in a floor are twenty-two feet long, three inches thick, twelve inches deep, and twelve inches from centre to centre. How much load per square foot will be required to break them if $R = 8,000$ pounds?

CHAPTER II.

KING-POST AND QUEEN-POST SYSTEMS.

KING-POST SYSTEM.

35.—THE KING-POST TRUSS is frequently employed in bridges of short span, and in the construction of roofs. Its

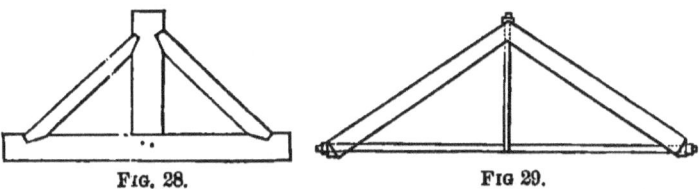

FIG. 28. FIG 29.

construction is simple, and it is very rigid. The common form is that shown in Fig. 28. In modern constructions a vertical iron tie is substituted for the post, and a horizontal iron tie is often substituted for the wooden one, as shown in Fig. 29. In bridges the inclined pieces are called braces, but in roofs they are called rafters. We consider two general cases: 1st, that in which the braces are equally inclined; and 2d, that in which they are unequally inclined.

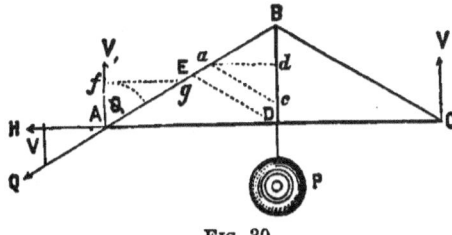

FIG. 30.

In bridges the load will be upon the supported chord AC, Fig. 30, and we may assume without being far from the truth that one-half the load on AD is supported at D, and the other half at A; and similarly on DC. This hypothesis is exact if there be a joint at D. But if the beam be continuous from A to C, and the load uniformly distributed, the point D will sustain $\frac{5}{8}$ of the total load (see Article 98, *Resistance of Materials*).

Let P be the load which is sustained at D. This is the only load which causes strains on the tie BD and on the braces; the remainder of the load being carried directly by the abutments. The load P produces a pull equal to P on DB, and thence a push on each of the braces AB and BC. Of the load P, each of the supports A and C carries $V = \frac{1}{2} P$.

Let $P =$ the strain on DB,

$V =$ the reaction of the support A due to P,
$H =$ the horizontal strain on the tie AC, due to P,
$Q =$ the strain on each of the braces,
$\theta =$ the inclination of the brace AB to the vertical,
$D = BD =$ the depth of the truss,
$l = AD = DC$, and
$L = AC =$ the span.

Suppose that the parts are reduced to rigid right lines, and that the joints at A, B, and C are perfectly flexible.

Take any convenient distance on the vertical line AV, to represent the reaction of V. Through the upper end of the line thus assumed, draw a line parallel to AD and prolong it till it meets AB;—and this line will represent the horizontal pull on the tie AC, and the distance from A to the point where the line intersects AB will represent the push on AB. Hence from this small triangle we have

$$\left. \begin{array}{l} V = \frac{1}{2} P \\ Q = V \sec \theta = \frac{1}{2} P \sec \theta \\ H = V \tan \theta = \frac{1}{2} P \tan \theta \end{array} \right\} \dots\dots (68)$$

We also have $\tan \theta = \dfrac{AD}{BD} = \dfrac{l}{D}$, and

$$\sec \theta = \frac{AB}{BD} = \frac{\sqrt{l^2 + D^2}}{D}$$

$$\left. \begin{array}{l} \therefore Q = \frac{1}{2} P \dfrac{\sqrt{l^2 + D^2}}{D} \\ H = \frac{1}{2} P \dfrac{l}{D} \end{array} \right\} \dots\dots (69)$$

KING-POST.

If the angle at B is right, θ will be 45 degrees and $l = D$;

$$\therefore \begin{array}{c} Q = \tfrac{1}{2}\sqrt{2}.\,P \\ H = \tfrac{1}{2}\,P \end{array} \Bigg\} \quad \ldots\ldots\ldots(70)$$

This may also be solved by directly resolving P, thus:—take any portion of DB, as Bc, to represent P, and draw ca parallel to BC. Then will Bac represent the triangle of forces; Ba representing the strain on BA, ac that on BC, and Bc that on BD. In this case, the rafters being equally inclined, $Ba = ac$. Draw ad parallel to AD and it will represent the stress on $AC = H$; and $Bd = dc = \tfrac{1}{2} P$. Hence we have from these triangles $ad = Bd\,tang\,\theta$, or $H = \tfrac{1}{2} P\,tang\,\theta$; also $aB = Bd\,sec\,\theta$ or $Q = \tfrac{1}{2} P\,sec\,\theta$, the same as before. We also have from similar triangles

$$aB : Bd :: AB : BD, \text{ or}$$
$$Q : \tfrac{1}{2} P :: \sqrt{l^2 + D^2} : D$$
$$\therefore Q = \tfrac{1}{2} P \frac{\sqrt{l^2 + D^2}}{D} \text{ as before.}$$

Having found the amount and character of the strains we proceed to proportion the parts. Observe that if there are two trusses to carry the load we divide the total strains by two. T being the modulus of the tenacity of the main tie, and K its section, we have

$$H = \tfrac{1}{2} P \frac{l}{D} = TK \therefore K = \frac{Pl}{2TD} \ldots\ldots(71)$$

For the vertical tie we have

$$K = \frac{P}{T} \ldots\ldots\ldots(72)$$

For the rafters or main braces, suppose that the ends are rounded, and we have, if their length exceeds fifteen times their diameter, (see article 24,)

$$Q = 13\,\frac{d^{3.76} - d_1^{3.76}}{l^{1.7}}, \quad \ldots\ldots\ldots(73)$$

for hollow cast-iron pillars; and if other materials or forms are used, the proper values can be taken from the table. If the length is less than fifteen times the diameter, use equation (67). Or use Gordon's formulas (article 22).

EXAMPLES.—1. If the span is 80 feet (which is too long for a bridge of this kind, but may do for a roof), depth 16 feet, uniform load one ton per foot of length, required the pull on the horizontal tie, and the push (or compression) on the rafters. Also required the proper dimensions of the rafters if they are square and made of dry deal. (*Note.*—In this problem assume that P is one-half the total load.)

2. If the span is 30 feet, the inclination of the rafters to the horizontal, 30°, the load on the central vertical tie, 10,000 lbs., required the depth of the truss and the strains upon the several parts.

3. Required the slope of the main brace so that the strain on it shall equal P. Find the corresponding strain on the main tie.

4. If the span is 40 feet, length of post 8 feet, $P = 2,000$ lbs.; required the strain on the braces, and thrust on the horizontal tie.

Ans. $Q = 2,692$ lbs. $H = 2,500$ lbs.

In the case of roofs, if the load is upon the main rafters, there will be no strain on the vertical tie due to that load, and its only service will be to support the long tie at the middle.

One of the objections to a flat truss of this kind, or, in other words, one in which the slope of the rafters is small, and in which the tie is composed of wood, is the difficulty of securing the ends so that they will not split out, or fail from *longitudinal shearing*. Fig. 31 shows some of the modes of securing them.

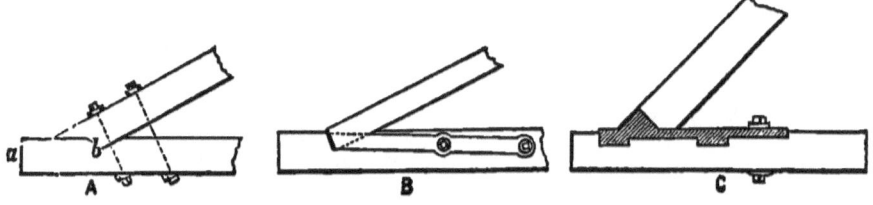

FIG. 31.

If $a\ b$ resists the thrust of the rafter by its *longitudinal shearing*, its length must be

$$a\ b = \frac{Hor.\ thrust}{50 \times breadth,}$$

for pine, hemlock, and spruce. For oak we would have

$$a\ b = \frac{Hor.\ thrust}{100 \times breadth.}$$

These are for an eight-fold security.

If the rafters are of wood, and the long tie of iron, it is easy to secure the ends. For the braces may be made to rest squarely against a cast-iron block, or shoe, to which the tie-rods may be attached in any convenient way. The tie may terminate with an eye for receiving a pin, and by the arrangement shown in Fig. 32, the cast-iron piece will be subjected to compression only.

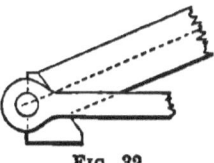

FIG. 32.

36.—INVERTED KING-POST.—If the king-post truss be inverted, and supported as before, at the ends of the long tie, as in Fig. 33, the amount of strain on each part will be the same as before, and hence may be computed by the same formulas; but the character of the strains will all be reversed. Thus, if a load is at D, the piece $D B$ will be compressed, the inclined pieces $A B$ and $B C$ will be under tension, and the piece $A C$ will be compressed.

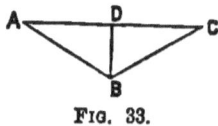

FIG. 33.

37.—A BRACED BEAM.—The middle of a beam may be supported by two braces, the lower ends of which rest against walls as in Fig. 34. In this case the strain on the braces is the same as in the king-post truss, and the thrust against the walls is the same as it would be upon a tie which would connect the feet of the braces. The pressure of the rafters against each other at their upper ends is equal to H. The beam which supports the load is not involved in this system of strains. It is a beam supported at three points, or of two beams supported at their ends.

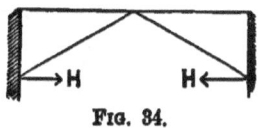

FIG. 34.

38.—MINIMUM VOLUME OF MATERIAL.—Given the length of the truss, and the load upon the vertical tie, it is required to find the depth of the truss, so that the total volume of material in the truss shall be a minimum.

Let D (fig. 29) be the unknown depth, and l half of the known span. K, the section of the vertical tie, may be found from the formula,

$$K = \frac{P}{T}$$

Letting the long dimensions (D and l) be in feet, and the transverse dimensions (K and d) be in inches, so as to conform with the notation in article 24, and we have for the volume in inches

$$12\,DK = \frac{12\,DP}{T} \quad \ldots \ldots \ldots \ldots \ldots \ldots (a)$$

The strain on the long tie is

$$\tfrac{1}{2}P \cdot \tang \theta = P\frac{l}{2D};$$

and its section is

$$K_1 = \frac{Pl}{2\,DT};$$

and its volume is

$$K_1 .24\, l = \frac{12\,Pl^2}{DT} \quad \ldots \ldots \ldots \ldots \ldots (b)$$

The strain on the brace is

$$\tfrac{1}{2}P\,\sec\theta = \tfrac{1}{2}P\frac{\sqrt{l^2+D^2}}{D},$$

hence if the braces are dry deal and square we have (see article 24)

$$\tfrac{1}{2}P\,\frac{\sqrt{l^2+D^2}}{D} = 7.81\frac{d^4}{l_1^2} = 7.81\frac{d^4}{l^2+D^2}$$

By solving this in reference to d^2, we have for the section of a brace in inches,

$$d^2 = \sqrt{\frac{P}{15.62\,D}\left(l^2+D^2\right)^{\frac{3}{2}}}$$

and the volume of both braces is

$$2 \times 12\sqrt{l^2+D^2} \times d^2 = 24\sqrt{\frac{P}{15.62\,D}\left(l^2+D^2\right)^{\frac{5}{4}}} \ldots \ldots (c)$$

Hence, by adding equations (a), (b) and (c) we have for the total volume in inches

$$\frac{12\,DP}{T} + \frac{12\,Pl^2}{T} + 12\sqrt{\frac{P}{3.905\,D}}\left(l^2+D^2\right)^{\frac{5}{4}}$$

which is to be a minimum. Hence by differentiating we have,

$$\frac{P}{T} - \frac{Pl^2}{TD^2} + \sqrt{\frac{P}{15.62}}\left(4\,D^2 - l^2\right)\left[\frac{l^2+D^2}{D^5}\right]^{\frac{1}{4}} = 0$$

The general value of D cannot be found from this equation, and hence the problem cannot be easily discussed.

EXAMPLE. 1. If $T = 12{,}000$ lbs., $l = 20$ feet, and $P = 20{,}000$ pounds; required D, and consequently the slope of the rafter for a minimum quantity of material.

2. If $T = 12{,}000$ lbs., $l = 30$ feet, and $P = 2{,}000$ pounds, required D.

QUESTION.—Does D increase or decrease as P increases?

If the truss be **INVERTED**, as in Fig. 33, and the notation the same as in the preceding example, and the load on the supported chord as before, we have for the section of the vertical post, if it be of oak (see article 24),

$$k = d^2 = D\sqrt{\frac{P}{10.95}},$$

and its

$$volume = D^2\sqrt{\frac{P}{10.95}} \dots \dots (f)$$

The strain upon the long chord will be $\dfrac{Pl}{2D}$, and its *section*, if of oak, will be $d^2 = l\sqrt{\dfrac{Pl}{21.9D}}$, and the volume of the whole chord will be,

$$volume = 2\,l^2\sqrt{\frac{Pl}{21.9\,D}} \dots \dots (g)$$

The strain upon the long ties will be $\tfrac{1}{2}P\dfrac{\sqrt{l^2 + D^2}}{D}$, and hence the section $= \dfrac{P}{2T}\dfrac{\sqrt{l^2 + D^2}}{D}$, and the volume of both ties will be,

$$volume = \frac{P}{T}\frac{(l^2 + D^2)}{D} \dots \dots (h)$$

∴ Total volume will be $D^2\sqrt{\dfrac{P}{10.95}} + 2\,l^2\sqrt{\dfrac{Pl}{21.9D}} + \dfrac{P}{T}\dfrac{(l^2+D^2)}{D}$,

which is to be a minimum.

$$\therefore \sqrt{\frac{P}{2.74}}D - \sqrt{\frac{Pl^3}{21.9}}\cdot\frac{1}{D^2} - \frac{Pl^2}{T}\cdot\frac{1}{D^2} + \frac{P}{T} = 0$$

from which the value of D may be found, and hence the slope of the tie-rods.

The problem is **SIMPLIFIED** by supposing that the resistance to compression varies directly as the section (which is true for short braces). For these we shall have for Fig. 29, the strain on the long tie as before $= \dfrac{P \sqrt{l^2 + D^2}}{2 D}$, and its section will be this value divided by C (modulus for crushing) and the volume of both braces will be this result multiplied by their length.

$$\therefore \text{volume of braces} = \dfrac{P (l^2 + D^2)}{C. D}, \dots\dots\dots (i)$$

which value added to equations (a) and (b) gives for *total volume*

$$\text{in the truss} = \dfrac{D P}{T} + \dfrac{P l^2}{D T} + \dfrac{P (l^2 + D^2)}{C D}. \dots\dots (j)$$

Hence,
for a minimum we have

$$\dfrac{P}{T} - \dfrac{P l^2}{T D^2} - \dfrac{P l^2}{C D^2} + \dfrac{P}{C} = 0,$$

Hence by factoring we find that

$$D = l,$$

consequently, the inclination of the braces will be 45 degrees. The other value given by the equation is inadmissible.

Hence the total volume of the truss becomes

$$2 P D \left(\dfrac{1}{T} + \dfrac{1}{C}\right) \dots\dots\dots (k)$$

If $T = C$ we have

$$\dfrac{4 P D}{T} = \dfrac{2 P L}{T};$$

that is, it is equivalent to a piece whose length is twice the span, and whose section is that required to sustain P by a direct pull.

The minimum material for the braces only is (using Eq. (i))

$$\dfrac{2 P D}{C} = \dfrac{P L}{C} \dots\dots\dots (l)$$

39.—MINIMUM DEPRESSION.—In a triangular frame, like Fig. 35, in which the braces and tie are all of the same material and uniform section, it is required to find the inclination of the braces, so that the depression of the vertex shall be a minimum for a load P, all the pieces being elastic.

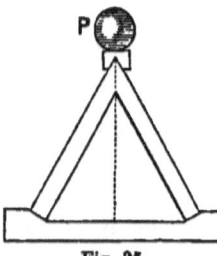

Fig. 35.

Let $P =$ the load at the vertex.
$l =$ the length of the base.
$a =$ the length of each brace.

D = the altitude of the triangle.
i = the angle which the brace makes with the horizontal tie.
K = the section of the pieces, and
E = the coefficient of elasticity.

Then $D = \sqrt{a^2 - \frac{1}{4}l^2}$, and by differentiation,

$$dD = \frac{ada - \frac{1}{4}ldl}{D} \quad \ldots \ldots \ldots \ldots (a)$$

in which dD is a small depression of the vertex, and is to be a *minimum*, da is a correspondingly small compression of a brace, and dl a correspondingly small extension of the tie.

We also have $a = \dfrac{l}{2\cos i}$; $D = \frac{1}{2} l \tan i$; strain on rafters $= \dfrac{P}{2 \sin i} = P'$; and tension on the tie $= \frac{1}{2} P \cot i$.

For the amount of compression, we have (see *Resistance of Materials*),

$$da = \frac{P'a}{EK} = \frac{P}{2\,EK\,\sin i}\frac{l}{2\,\cos i} = \frac{Pl}{4\,EK} \cdot \frac{1}{\cos i \, \sin i};$$

Similarly,—$dl = \dfrac{Pl \cot i}{2\,EK}$

Hence, equation (a) becomes

$$dD = \frac{Pl}{4\,EK}\frac{1}{\sin^2 i}\left[\frac{1}{\cos i} + \cos^3 i\right]$$

which is to be a minimum. By differentiating, placing equal to zero, and reducing, we have

$$2\cos^5 i + 3\cos^3 i = 1,$$

one of the roots of which is

$$\cos i = \tfrac{1}{2}$$
$$\therefore i = 60°;$$

hence the triangle must be equilateral.

40.—TRUSSED BEAMS.—In a trussed beam, like Fig. 36, the total compression on the upper side is that due to the bending of the beam added to the horizontal pull of the truss rods; and the tension on the lower side is that due only to the bending of the beam. But it is difficult to determine the values of these strains, for the strain on each is dependent

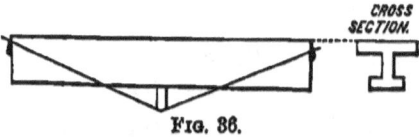

Fig. 36.

upon the distortions (compressions or extensions) of the other. So far as cast-iron beams are concerned, the value of such trussing has been determined experimentally by Wm. Fairbairn, the results of which are reported in his work on *Cast and Wrought Iron*. The beams experimented upon were of the double **T** form, as shown in Fig. 36, and the truss rods were of wrought iron. He found it exceedingly difficult to adjust the strains upon the iron rods so as to secure the best result, but concluded that they should be so adjusted as to secure a strain of 2 or 3 tons before the permanent load was placed upon the beam. He also concluded that such beams were not reliable.

The trussing of cast-iron beams, however, is not very important at the present day, for solid rolled wrought iron ones can be made quite as cheaply, and are much more reliable.

41.—RAISED TIE, OR DOUBLE RAFTERS.—If the lower chord (or tie) be raised, as in Fig. 37, the strains are considerably modified. This form of truss is common in roof construction, in which case the parts AD and BD are prolonged to the main rafters, as shown by the dotted lines, and are secured to them. In the case of roofs, the load will be distributed evenly over the main rafters, in which case the strains upon AD, DB, and DC, will be the same as if one-half the load were concentrated at C, and the other half sustained directly by the abutments. In order, therefore, to

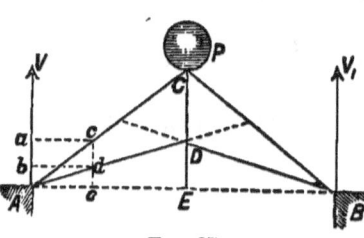

Fig. 37.

simplify the problem, we will consider the case in which a load P is placed at the vertex C, and the joints at D and C are perfectly flexible. The pieces AC and BC will be compressed; AD, DB, and DC will be subjected to a *pull*. At the joint D, there will be an upward pull equal to the tension upon CD, and at C there will be a downward pull equal to the tension on CD added to the weight P. It is also evident that the horizontal *push* outward of AC will be equal to the inward horizontal *pull* of AD. In other words, the horizontal component of the strains in AC and AD neutralize each other. This view of the case makes the solution very simple, for we may suppose that the truss is divided into two trusses; one of which will be ACB, with a horizontal chord AB; and the other ADB, with the same horizontal chord.

Let $P =$ the load at C,

$t =$ the tension on CD,

$H =$ the horizontal component of the push and pull,

$t_1 =$ the tension on AD, and

$Q =$ the compression on AC.

Take any distance Ae on the chord AB, to represent the horizontal component, H, of the strains; erect a perpendicular ec, and from d and c, where it intersects the rafters, draw the horizontals ca and db, then will Aa represent the reaction of the support for the truss ACB, and Ab the reaction for the truss ADB; Aa being a *push* up, and Ab being a pull down, their difference, ab, will be the amount which is actually sustained by the abutment, and is $\frac{1}{2} P$. Ad represents the pull on AD, and Ac the push on AC.

For the truss ADB we have

$$Ae \; : \; ed \; :: \; AE \; : \; ED, \text{ or}$$
$$H \; : \; \tfrac{1}{2} t \; :: \; AE \; : \; ED$$
$$\therefore H = \tfrac{1}{2} t \, \frac{AE}{ED} \ldots\ldots\ldots\ldots\ldots\ldots\ldots(74)$$

For the truss ACB we have

$$Ae : ec \; :: \; AE \; : \; EC, \text{ or}$$
$$H \; : \; \tfrac{1}{2}(P + t) \; :: \; AE \; : \; EC.$$
$$\therefore H = \tfrac{1}{2}(P + t) \frac{AE}{EC} \ldots\ldots\ldots\ldots(75)$$

Equations (74) and (75), being equal, give

$$t\frac{AE}{ED} = (P+t)\frac{AE}{EC}$$

$$\therefore t = \frac{ED}{DC}P \dots \dots \dots \dots (76)$$

which is independent of the slope of the rafters.

We also have $ed : Ad :: ED : AD$, or

$$\tfrac{1}{2}t : t_1 :: ED : AD$$

$$\therefore t_1 = \frac{t}{2}\frac{AD}{ED} = \frac{AD}{2DC}P \dots \dots \dots (77)$$

Also,

$$ec : Ac :: CE : AC;\text{ or}$$
$$\tfrac{1}{2}(P+t) : Q :: CE : AC$$

$$\therefore Q = \frac{AC}{2CE}(P+t) = \frac{AC}{2DC}P \dots \dots (78)$$

DISCUSSION.—From Eq. (76) we have, if $ED = DC$, $t = P$. If $DC = 0$, $t = \infty$. If $ED = 0$, $t = 0$.

From Eq. (77), if $DC = CE$, $t_1 = \dfrac{AE}{2CE}P = \dfrac{l}{2D}P$, which is the same as the second of Eqs. (69).

From Eq. (78), if $DC = 0$, $Q = \infty$. If $DC = EC$, $Q = \dfrac{AC}{2EC}P$, which is the same as the first of Eqs. (69).

Next suppose that the load is placed at D, as in Fig. 38. Then will the vertical stress at the joint D be $P - t$; and at the joint C it will be t. Proceeding as before and we have

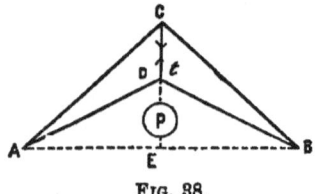

Fig. 38.

$$\frac{P-t}{ED} = \frac{t}{EC} \therefore t = \frac{EC}{ED + EC}P \dots \dots (79)$$

If $ED = 0$, $t = P$, as it should.

42.—DEPRESSED TIE.—If the lower tie be depressed below a horizontal, as in Fig. 39, the corresponding joints being lettered the same as in the preceding cases, the same forms of expression will give the strain on the several parts. Thus, if the load be placed at C, the compression on CD will be

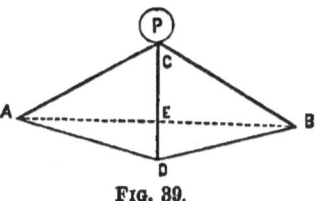

FIG. 39.

$$t = \frac{ED}{DC} P,$$

the same as Eq. 76.

If the load be at D, the expression for the tension will be

$$t = \frac{EC}{ED + EC} P = \frac{EC}{DC} P$$

the same as Eq. (79). These expressions become identical when $ED = EC$. They are also independent of the slope of the rafters.*

In these cases if the load be uniformly distributed over the upper or the lower rafter, all the strains will be as above, except that on the rafter; and the strains on that member may be found as in Article 20.

43.—SOLUTIONS BY DIAGRAMS.—To solve the case of the simple king-post, Fig. 30; draw a straight line CE, Fig. 40, in the direction of the acting load (vertical), then take any point O as an origin, and draw OC, OD, and OE respectively parallel to AB, AD, and BC; then will $CE = P$; $CD = \frac{1}{2}P = DE$; $OD = H$, the horizontal thrust, and $OC = Q =$ the compression on the rafter. From this figure we have

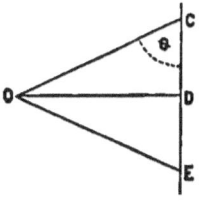

FIG. 40.

$$OD = H = \tfrac{1}{2} P \tan\theta$$
$$OC = Q = \tfrac{1}{2} P \sec\theta$$

which are the same as equation (68).

---

* See several analytical solutions of the preceding problem, by S. W Robinson, Class of 1863, Univ. of Mich.—*Jour. Frank. Inst.* 3d Series, p. 13

In a similar way to solve the case shown by Fig. 37, draw a line CG, Fig. 41, take a point O and draw OC parallel to AC; OD parallel to AD; OF parallel to DB; OG parallel to CB. Then will $CD + FG = P$; $DF = t$; $OC = Q$; $OD = t_1$.*

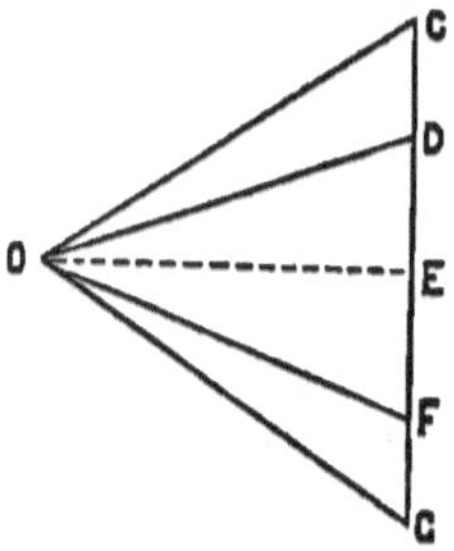

Hence we have directly from this figure $\frac{1}{2}, t : \frac{1}{2} P :: ED : DC \therefore t = \frac{ED}{DC} P$, as before, and similarly for the others.

FIG. 41.

44.—FINK TRUSS.—A *Fink Truss* consists of a combination of *king-posts* with equally inclined ties, as in Fig. 42. It has the primary system ACB; the two secondary systems AhD

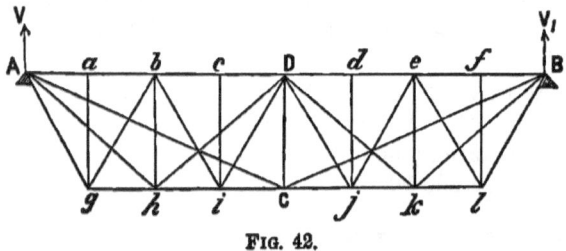

FIG. 42.

and DkB; the four tertiary systems Agb, biD, Dje, and elB, and so on, when there are more systems.

The load may be upon the upper or lower chord, but if it be upon the upper chord, there will be no necessity for a lower chord. Suppose that the load is uniformly distributed over the whole length of the supported chord. The divisions Aa, ab, &c., are called *bays*.

Let W = the total load,
L = the span = AB,
N = the number of bays in the supported chord,
$l = L \div N$ = the length of each bay,

---

* For a demonstration of the principle upon which this solution is founded see Appendix.

FINK TRUSS.

$D =$ the depth of the truss, and
$P =$ the load on a joint.

$$\therefore P = \frac{W}{N}.$$

Suppose that a weight P is placed on each of the joints a, b, c, &c. Then the joint D carries $\frac{1}{2} P$ which is at c on the truss $b\ i\ D$, and $\frac{1}{2} P$ which is at b on the truss $A\ h\ D$, and so on. Hence D carries one-half the total load, or $4\ P$ in this case. Similarly, b and e each carry one-fourth the total load, or $2\ P$ in this case; and a, c, d, and f each carry P in this case. These values are the strains on the vertical struts.

The first of equations (69) gives the strains on the ties. We have

Strain on $A\ g = \dfrac{\sqrt{l^2 + D^2}}{2\ D} P =$ strain on $b\ g$, $b\ i$, $i\ D$, $D\ j$, $j\ e$, and $e\ l$.

Strain on $A\ h = \dfrac{\sqrt{(2l)^2 + D^2}}{2\ D} 2\ P = \dfrac{\sqrt{4\ l^2 + D^2}}{D} P =$ strain on $h\ D$, $D\ k$, and $k\ B$.

Strain on $A\ C = \dfrac{\sqrt{(4l)^2 + D^2}}{2\ D} 4\ P = \dfrac{\sqrt{16\ l^2 + D^2}}{D} 2\ P =$ strain on $C\ B$.

The horizontal strain in the supported chord is uniform throughout, and equal to the sum of the strains due to each truss.

Total compression on supported chord $= \frac{1}{2} P \dfrac{l}{D} + \frac{1}{2} . 2\ P \dfrac{2\ l}{D} + \frac{1}{2} . 4\ P \dfrac{4\ l}{D} = \frac{1}{2} P \dfrac{l}{D} [1 + 4 + 16 + $ &c., if there were more terms.$]$

$$= 10\tfrac{1}{2} P \dfrac{l}{D} \text{ in this case.}$$

EXAMPLE.—Suppose that a *Fink Truss* is 96 feet long, and is divided into 8 equal *bays*, and the depth of the truss is 16 feet, and is loaded with 120 tons uniformly distributed, including its own weight. It is required to find the strains upon all the ties and braces.

45.—ROOF TRUSSES, which somewhat resemble the Fink Truss, as shown in Fig. 43, are somewhat common. In this case, the lower ends of the single king-post trusses are in the outline of the main truss. In a roof, this truss is placed in an inclined position, and sustains a vertical load (chiefly), in which cases the analysis of strains differs considerably from the preceding, as will be seen hereafter. If the truss is horizontal, the mode of analysis is evident from the preceding problem.

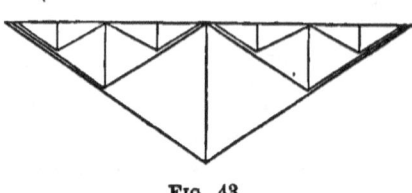

FIG. 43.

46.—UNEQUALLY INCLINED BRACES or TIES.—If the vertex B of the truss is not over the middle of $A\ C$, the braces $A\ B$ and $B\ C$ will be unequally inclined. Whatever be the character of the loading, suppose that the strain on the tie $B\ D$ is determined.

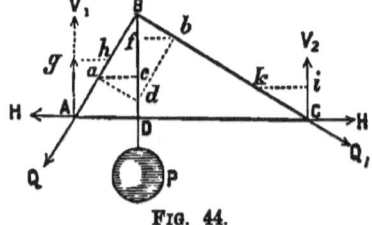

FIG. 44.

Let $P =$ the part of the load which is supported at D,
$Q =$ the strain on $A\ B$,
$Q_1 =$ "   "   " $B\ C$,
$H =$ "   "   " $A\ C$,
$V_1 =$ the amount which A sustains due to the loading at D (and does not include that part of the uniform load which A sustains directly),
$V_2 =$ the reaction at C,
$i =$ the angle $B\ A\ D$,
$i_1 =$ "   " $B\ C\ A$,
$\theta =$ "   " $B\ A\ V_1$,
$l =$ the part $A\ D$,
$l_1 =$ "   " $D\ C$, and
$L = l + l_1 = A\ C =$ the total span.

UNEQUALLY INCLINED RAFTERS.

By the principle of the lever we have,

$$V_1 L = P l_1$$
$$\therefore V_1 = \frac{l_1}{L} P.$$

Similarly, $V_2 = \frac{l}{L} P.$

Take any distance $A g$ to represent V_1, and draw $g h$ parallel to $A D$, and $g h$ will represent the strain on $A D$, and $A h$ the strain on $A B$. Hence we have

$$H = V_1 \, tang \, \theta = \frac{l_1}{L} P \cdot \frac{l}{D} = \frac{l \cdot l_1}{L \cdot D} P \dots \dots (80)$$

$$Q = V_1 \, sec \, \theta = \frac{l_1}{L} \frac{\sqrt{l^2 + D^2}}{D} P \dots \dots (81)$$

Similarly, $Q_1 = \frac{l}{L} \frac{\sqrt{l_1^2 + D^2}}{D} P \dots \dots \dots (82)$

SECOND SOLUTION. Take Bd to represent P. Draw ad parallel to BC, cd parallel to AB, and ac parallel to AD. Then

$$Q : P :: Ba : Bd$$
$$\therefore Q = \frac{Ba}{Bd} P.$$

Also $Bc : ac :: DB : AD$; $ac : cd :: CD : BD$, and $ac : AD :: aB : AB$

By combining these we have $\frac{Ba}{Bd} = \frac{BA \cdot CD}{AC \cdot BD}$

$$\therefore Q = \frac{BA \cdot CD}{AC \cdot BD} P = \frac{l_1 \sqrt{l^2 + D^2}}{LD} P \dots \dots \dots (83)$$

which is the same as Eq. (81).

$$Q : H :: Ba : ac :: BA : AD :: \sqrt{l^2 + D^2} : l$$
$$\therefore H = \frac{AD \cdot DC}{AC \cdot BD} P = \frac{l \cdot l_1}{L \cdot D} P, \text{ as before} \dots \dots \dots (84)$$

THIRD SOLUTION. $Ba : Bd :: \sin a d B : \sin B a d,$
or $Q : P :: \cos i_1 : \sin (i + i_1)$

$$\therefore Q = \frac{\cos i_1}{\sin(i + i_1)} P \dots \dots \dots (85)$$

Similarly,

$$Q_1 = \frac{\cos i}{\sin (i + i_1)} P \dots \dots \dots (86)$$

Also

$$H = \frac{\cos i \cos i_1}{\sin (i + i_1)} P \dots \dots \dots (87)$$

In these equations if $l = l_1$ or $i = i_1$ this case reduces that of *equally inclined* rafters, and the formulas become the same as for that case.

If the angle at B is right, $i + i_1 = 90$; and equations (85), (86) and (87) become

$$Q = P \cos i_1 = P \sin i. \quad \ldots \ldots (88)$$
$$Q = P \cos i = P \sin i_1. \quad \ldots \ldots (89)$$
$$H = P \cos i \cos i_1 = P \cos i \sin i. \quad \ldots \ldots (90)$$

If the braces are equally inclined, $i = i_1$ and (85), (86) and (87) become

$$Q = Q_1 = \frac{\cos i}{\sin 2i} P = \tfrac{1}{2} \frac{P}{\sin i} = \tfrac{1}{2} P \operatorname{cosec} i.$$

And $H = \dfrac{\cos^2 i}{\sin 2 i} P = \tfrac{1}{2} P \dfrac{\cos i}{\sin i} = \tfrac{1}{2} P \cot i.$

FOURTH SOLUTION. In the figure we have

$Bc : ca :: BD : DA$ or $Bc : H :: D : l$
$dc : ca :: Bf : fb :: BD . DC$ or $cd : H :: D : l_1$
$Bc + cd = P$

$\therefore H = \dfrac{l l_1}{L.D} P$, as before.

47.—If the truss be **INVERTED**, as in Fig. 45, the supported chord becomes a *straining* beam, and the inclined pieces *ties*. If the load be upon the supported chord at D, the piece DB becomes a strut, and directly supports P; but if P is suspended at B the only office of DB will be to support the chord at D. If the upper chord be inclined upward, as in Fig. 46, and

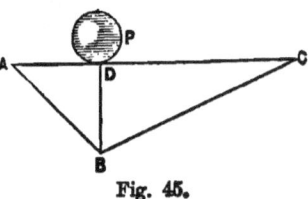

Fig. 45.

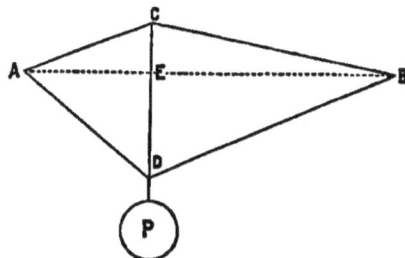

Fig. 46.

the load be suspended at D, the piece DC becomes a tie, and the compression upon it will be

$$t = \frac{EC}{DC} P$$

If P be placed at C, we have

$$t = \frac{ED}{DC}P,$$

and hence if DC exceeds ED the compression on CD will be less than P; but if the point C falls below E, and the load remains at C, the compression on CD will exceed P.

48.—BOLLMAN'S TRUSS.—A skeleton or outline of Bollman's Truss is shown in Fig. 47. In this case, as in many others, the novel feature does not consist so much in the outline or skeleton as in the details of construction. Mechanically speaking, this truss consists of a series of king-posts with

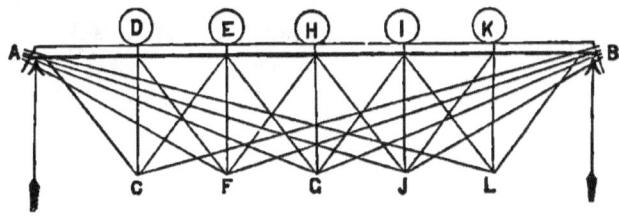

Fig. 47.

unequally inclined ties. The vertical pieces are struts, and the long chord is common to all the several trusses.

The details of the construction consist in the manner of joining the several parts. Thus, the chord, instead of being continuous, is made in sections, or separate pieces of hollow cast iron, which abut against each other at the joints D, E, H, etc., as shown in Fig. 48. They are usually octagonal. The long tie rods, AC, AF, etc., are not attached rigidly to the lower ends of the posts DC, EF, etc., but instead thereof each pair, as AC and CB, AF and FB, etc., is attached to a link, which link is attached to the lower ends of the posts. This prevents cross strains upon the posts when the rods are expanded or contracted by changes of temperature or changes of strains. The rods of a pair, as AC, CB, being of unequal length, will be unequally strained for any load (see Eqs. (81) and (82)), and should be proportioned to resist the strains. If they are so proportioned, they, being of unequal length, will be unequally elongated for any change of load, and hence would tend to carry the point C to the left. If they are contracted, they would tend to carry the point C to the right. So of other points. The link offers some flexibility at these points.

Fig. 48.

The short diagonals in each of the panels $DCFE$, $EFGH$, etc., Figs. 47 and 48, serve directly to keep the joints D, E, etc., in place; for if the chord is made convex upward, or if it becomes so relatively by the depression of some point, there is a tendency, due to the compression in the chord, to cause the convex part to rise upward, as indicated in Fig. (46). The panel rods will prevent such displacement. They may also serve for giving additional security; for if made sufficiently strong, they may transmit the strains to the other parts in case one of the long tie rods breaks. Thus, if FB is broken, the panel rods in the diagonals DF and FH will carry the load to the posts DC and HG. The succession of king-posts may be called the **PRIMARY SYSTEM**, and the system of posts and tie rods, which, as will be shown hereafter, may form a complete truss by itself, may be called the **SECONDARY SYSTEM**.

The long tie rods pass through slots at the ends of the chords, and are secured by a pin passing through an eye. A lower chord sometimes passes through the lower joint for supporting the roadway. In other cases the roadway rests upon the upper chord.

49.—ANALYSIS OF BOLLMAN'S TRUSS—PRIMARY SYSTEM.
—It will be observed in the *primary system* of this truss that the load at each joint is carried directly to the abutments at A and B, and hence we may determine the strains upon the tie rods and posts for a load upon any one of the joints. Or, in other words, the load at a joint affects only one set of tie rods and one post.

Let $l = AD = DE = EH$, etc., = the length of one bay, Fig. 47,

$D = DC =$ the depth of the truss,
$L = AB =$ the span,
$N =$ the number of bays in AB,
$n =$ the number of a bay considered, counting from either end. Thus, if AD be called the first bay; DE, the second, etc., HE may be called the n-th. This will also give the number of a tie counting from either end. Thus, if AC is the first, AF the second, etc., AJ will be the n-th.

$Q_1 =$ the strain on the first tie, or AC,
$Q_2 =$ the strain on the second tie, or AF,
$Q_n =$ the strain on the n-th tie,
$H_1 =$ the compression on the chord due to the first truss, or ACB,
$H_2 =$ the compression on the chord due to the second truss, or AFB,
$H_n =$ the compression on the chord due to the n-th truss.

In regard to the long ties it is necessary to consider only those which incline one way, for those which incline the same amount in the opposite direction will evidently receive the same stress for the same load.

For the strain upon the n-th tie for a load P upon any joint, as H, we may generalize Equation (81) or (83), and have

$$Q_n = \frac{(N-n)\,l\,\sqrt{n^2 l^2 + D^2}}{L.D} P \ldots\ldots\ldots (91)$$

Hence the strain upon any particular tie is found by giving to n its particular value in Equation (91). Thus, for AC make $n = 1$, for AF make $n = 2$, and so on. Hence,

The strain on 1st tie is $Q_1 = (N-1)\sqrt{l^2 + D^2}\dfrac{l.P}{L.D}\ldots(92)$

The strain on 2d tie is $Q_2 = (N-2) \sqrt{4\,l^2 + D^2}\, \dfrac{l.P}{L.D}$ ..(93)

The strain on 3d tie is, $Q_3 = (N-3) \sqrt{9\,l^2 + D^2}\, \dfrac{l.P}{L.D}$.(94)

In a similar way, for the strain upon the chord due the n-th truss, we generalize Eq. (84), and have

$$H_n = \frac{nl\,(N-n)\,l}{L.D}\,P$$

Hence the strain due to the 1st truss is $H_1 = (N-1)\,\dfrac{l^2}{L.D}\,P$

" " " " " 2d " " $H_2 = 2\,(N-2)\,\dfrac{l^2}{L.D}\,P$

" " " " " 3d " " $H_3 = 3\,(N-3)\,\dfrac{l^2}{L.D}\,P$

The total stress on the chord is

$H_1 + H_2 + H_3 +$ etc., to $H_{N-1} =$

$$\left[\begin{array}{l} N + 2\,N + 3\,N + \text{etc., to } N-1 \text{ terms} - (1 + 2^2 \\ + 3^2 + 4^2 + \text{etc., to } (N-1)^2) \end{array}\right] \frac{l^2}{L.D}\,P$$

$$= \frac{(N^2-1)\,l}{6\,D}P\dots\dots\dots\dots(95)$$

In ordinary cases the bridge will be supported by two trusses, and hence the strains found above must be divided by two to get the strains on each truss.

.The chord is so nearly horizontal that when the load is on the upper chord the strain on each post is P; but if the chord angles upward at any joint, the post under that joint will sustain less than P, but if it angles downward it will sustain more than P, as shown in article 47.

If the **SECONDARY SYSTEM** of truss rods is to insure safety, the strains on them may be computed by the first of Equations (69).

EXAMPLE. The following data are taken for a bridge as manufactured by Charles Kellogg & Co. (formerly in Detroit, Mich.)

$L = 100$ feet $=$ the span,
$N = 8 =$ the number of bays,
$\therefore l = 12\tfrac{1}{2} =$ the length of one bay,
$W_1 = 60$ tons (net) $=$ the weight of the frame,
$W_2 = 100$ tons $=$ the assumed uniform load,
$P = 20$ tons $=$ the maximum load at each post, and
$D = 19$ feet $=$ the depth.

BOLLMAN'S TRUSS.

Required the dimensions of the parts when one-fourth the statical stress is added for shocks. This addition is equivalent to calling P, 25 tons. Also consider the tenacity, per square inch, T, of wrought iron, equal to the compression, C, per square inch of cast iron, and call $T = C = 11,000$ lbs.

There were two trusses, and two rods in each truss for supporting the load P.

Hence by means of Eq. (91), and the above value of T, we have, counting from either end,

	Tons.	Section of each tie. Square inches.
Stress on the 1st tie	26.180	1.12
" " 2d "	30.986	1.41
" " 3d "	34.569	1.57
" " 4th " 35.198 (middle one)		1.63
" " 5th "	32.230	1.48
" " 6th "	25.450	1.15
" " 7th "	14.733	0.74

The compression on the upper chord will be, according to Eq. (95), 86.35 tons = 172,700 lbs.

If the sections were solid and octagonal, their diameter, in order to resist flexure with a twenty-fold security, will be, according to Eq. (58)

$$d = 2 \sqrt{\frac{172,700 \times 150^2}{0.87527 \times 16,000,00 \times 3.1416^2}} = 9\tfrac{1}{4} \text{ inches.}$$

They were 10 inches in diameter, and hollow. I do not know the thickness, but according to Francis' tables they should be about one inch thick for a five-fold security.

The *posts* each were to sustain 10 tons, and were six inches in diameter.

The *panel* rods were to sustain 10 tons, and hence, according to the first of Eqs. (70), the strain on each would be 14.10 tons for both trusses; and 7.475 tons for each. Hence the section of the panel tie should be 1.26 square inches.

CHAPTER III.

TRUSSES IN WHICH THE UPPER AND LOWER CHORDS ARE HORIZONTAL.

TRAPEZOIDAL TRUSS.

50.—The form of the **TRAPEZOIDAL TRUSS** is shown in

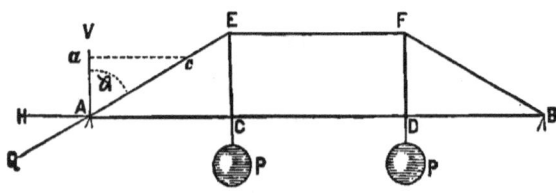

FIG. 49.

Fig. 49. All parts are most strained when it is loaded throughout, or when both joints C and D are loaded to their maximum. If the load is uniformly distributed over the lower chord, the joints C and D will each sustain one-third of the total load, the remaining third being sustained directly by the supports A and B. That is, A will sustain one-sixth of the load directly, and B the same, and each will also sustain one-third of the load as it is transmitted through CE and DF, and thence to A and B, through the braces AE and FB, thus making one-half the total load as it should. But only that part of the load which is supported at C and D produces strains on the trussing. Of this load the supports A and B each sustain an amount equal to P.

Let $l = AC = CD = DB$,
$D = EC$,
$V =$ the reaction at A,
$V_{\text{,}} =$ the reaction at B, and
$\theta =$ the angle VAE.

Th. in $V = P = V_1$

Strain on $AE = V \sec \theta = \dfrac{\sqrt{l^2 + D^2}}{D} P \ldots \ldots (96)$

Strain on $AC = V \tan \theta = \dfrac{l}{D} P \ldots \ldots \ldots (97)$

which equals the strain on $EF =$ that on $CD =$ that on DB
If the load is on the lower chord, the strain on the vertical ties
will be P, but if it be upon the upper
chord, as in Fig. 50, the vertical pieces
will simply support the lower chord,
and hence may, in such cases, be very
small.

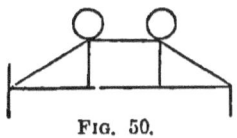
FIG. 50.

If only one joint is loaded, or when
the load on one joint somewhat exceeds the load on the other
joint, it may be said that the truss is unbalanced, and may
become distorted, as shown in Fig. 51.
For such a load the truss is not completely braced. There should be *braces*
or *ties* in the diagonals of the panel
$CDFE$, Fig 49.

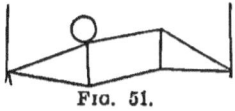
FIG. 51.

51.— TRUSS FULLY BRACED.—The preceding example
shows that a truss may be braced so as not to change its form
under a certain load, as in Fig. 49, but under other distributions
of the load it may become distorted. In such cases it is only
partially braced.

A TRUSS IN WHICH THE PARTS ARE CONSIDERED PERFECTLY RIGID
AND JOINTS PERFECTLY FLEXIBLE, IS FULLY BRACED (OR SIMPLY
BRACED) WHEN THE PARTS ARE SO ARRANGED THAT IT WILL NOT
CHANGE ITS FORM UNDER ANY ARRANGEMENT OF THE LOAD WHICH
IT IS INTENDED TO CARRY.

In practice the joints are not perfectly flexible and the parts
are elastic; and hence a change of form will take place for
every change of load, but when the conditions are fulfilled according to the preceding hypothesis, the frame will not become
distorted, as in Fig. 51; and it is this *distortion* that we wish to
prevent by bracing (or tieing). A truss may be distorted or
broken by an excessive load which shall endanger the strength
of the material, but this case is excluded by the definition. In

68   TREATISE ON BRIDGES.

other words, a truss is braced when the trussing prevents any turning about the joints beyond that which results from the yielding of the material on account of its elasticity.

52.— TRAPEZOIDAL TRUSS MODIFIED.—A beam is sometimes supported, as shown in Fig. 52, in which case the thrust of the braces is resisted at their lower ends by the walls. The same thing is shown in Fig. 53, and it may be distorted, as shown in Fig. 54, by a partial load.

53.— THE STRAINS upon the several parts of the *modified truss* may be found by the same formulas as given in the preceding case.

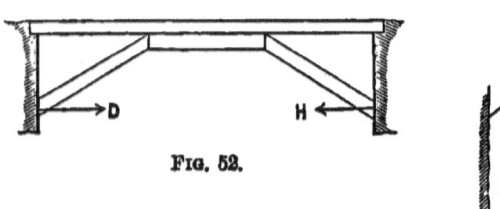

FIG. 52.

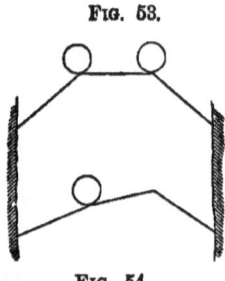

FIG. 53.

FIG. 54.

54.— TRAPEZOIDAL TRUSS INVERTED.—If the truss be inverted, as shown in Fig. 55, the *amount* of the strains on the several parts will be the same as in the erect truss, Fig. 49, but the character of all the strains will be changed; that is, the inclined pieces will here be ties instead of braces, the supported chord and vertical pieces will be compressed instead of extended. If the end braces are unequally inclined, the mode of solution will be essentially the same as in the preceding case.

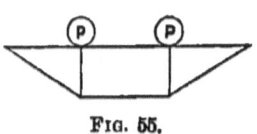

FIG. 55.

55.— EXAMPLES.—1. In a trapezoidal truss, suppose that the span is forty feet; length of posts eight feet, each bay one-third of the span; the strain on each post, 2,000 lbs.; required the strains on the braces and chords.

Ans. $Q = 3,887$ lbs.
$H = 3,333$ lbs.

Note.—Compare these results with the fourth example of article 35, observing that the truss in this example carries double the load of that in the former. If the load is uniformly distributed, it really carries more than double, for in each case they carry half the load between the ends and first joints, and in the case of the trapezoidal truss, the end bays are shorter than those in the king-post truss, thus throwing less load *directly* upon the abutments in the former than in the latter case; and hence requiring the posts or post to carry more.

2. If the span is eighty feet, depth of truss sixteen feet, uniform load one ton per foot of length; required the pull on the horizontal tie, and the compression on the braces and straining beam, there being three equal bays. Also required the proper dimensions of the braces if they are square and made of dry deal. (*Note.* $P = \frac{1}{2}$ the total uniform load.)

Compare the results with the 1st example in article 35.

TRIANGULAR TRUSS.

UPPER AND LOWER CHORDS HORIZONTAL.

56.—A **TRIANGULAR TRUSS** is one in which the space occupied by the truss is divided into triangles by the pieces which compose the truss, and in which each piece may act both as a tie and brace. Such a form is shown in Figs. 60, 61, and 62. A triangular truss, in which the ties and braces are equally inclined, is generally known as the **WARREN'S GIRDER**, or **WARREN'S TRUSS**, although English writers generally confine it to a triangular truss in which the ties and braces are inclined about 45 degrees to the vertical. The *Warren Girder*, as constructed in England, is usually made of iron. A *triangular wooden truss* was patented in this country by a Mr. Godard, and called *Godard's self-supporting truss.** In this truss the triangles were isosceles, and the sides inclined about 30 degrees to the vertical. The peculiarity consisted chiefly in the modes of fastening the ends of the pieces. Another *triangular truss* was invented in this country which is essentially a double lattice, as shown in Fig. 68, and which by way of distinction was called an Isometrical Truss. The Fink Triangular Truss is another form which is composed partly of wood and partly of iron.† Another truss, called Smith's Patent, ‡ originally composed entirely of wood, has some peculiar features, but which it is not necessary for our purpose to describe.

The analysis of all triangular trusses is essentially the same.

57.—**TRIANGLES THE ONLY PROPER FIGURES FOR TRUSS-WORK.**—The triangle is the only geometrical figure in which the angles cannot be changed without changing the lengths of the sides. Hence, to form a truss which will not distort when partially loaded, the truss-work should form tri-

---

* See *Appleton's Mechanic's Magazine*, May, 1852, p. 141.
† Manufactured by the Louisville Bridge and Iron Company.
‡ Manufactured by R. W. Smith, of Toledo, Ohio.

angular figures. But it does not follow from this that every such truss is *completely* braced. If the inclined pieces serve only as braces, as in Fig. 75, it may become distorted for a partial load, as in Fig. 76. Similarly, if the inclined pieces serve only as ties, it will be only *partially* braced (tied) for certain partial loads, and the distortion will be of the same kind as in Fig. 76. In all such cases it is necessary to have braces inclined both ways, as in Fig. 77, or ties both ways, as in Fig. 79. It will be seen from this that a truss is not *triangular* simply because it is composed of triangles, but it is essential that the inclined pieces be **TIE-BRACES**.

58.—The following **CONDITIONS WILL BE ASSUMED** in making an analysis of trusses. The parts will be considered as reduced to rigid right lines. The meeting of two or more lines at their common point of action will be called a **JOINT**. The *joints* will be considered—for the present at least—as perfectly flexible.

The strain will be considered uniform between adjacent joints; also that the same piece cannot be subjected to tension and compression at the same time, but that the resultant strain is the algebraic sum of the two. Tension will be called +, and compression −.

59.— **NOTATION.**—As far as practicable, the following notation will be common to all the cases:—

$P =$ the load applied at a *single* joint when there is but one weight;

$p =$ the weight at a single joint when several joints are loaded, and the several weights are equal to each other;

p_1, p_2, etc., = the weights when they are unequal;

$L =$ the length of the span of a truss,

$l =$ the length of a bay,

$D =$ the uniform depth of a truss when the chords are parallel, and the *greatest* depth if they are not parallel;

$\theta =$ the inclination of a brace or tie to the vertical;

$i =$ the inclination of a brace or tie to the horizontal;

$w =$ the weight per foot of length of the *dead* load when it is uniform;

w' = the weight per foot of length of the *live load* when it is uniform;

δ = the weight of a unit of volume of the material considered;

W_1 = the total weight of the truss;

W_2 = the total weight of the surcharge;

W = the total load on the structure = $W_1 + W_2$;

V_1 = the reaction of the support at one end;

V_2 = the reaction of the support at the other end;

C = the modulus of compression;

T = the modulus of tenacity;

K_n' = the section of the n-th bay of the lower chord;

K_n = the section of the n-th bay of the upper chord;

k_n = the section of the n-th tie or brace;

c_n = the compression of the n-th bay of the compressed chord;

t_n = the tension of the n-th bay of the tensioned chord; and

Σ is used to denote the summation of similar quantities.

Other notation will be given as it is needed.

60.—CASE I.—A SEMI-TRUSS HAS A SINGLE WEIGHT, P, AT ITS FREE END; IT IS REQUIRED TO FIND THE STRAINS UPON THE TIES, BRACES, AND CHORDS; THE BRACES AND TIES BEING EQUALLY INCLINED.

First, Consider a geometrical solution. Take a vertical line ab, to represent P, and draw bc parallel to ak, and prolong it

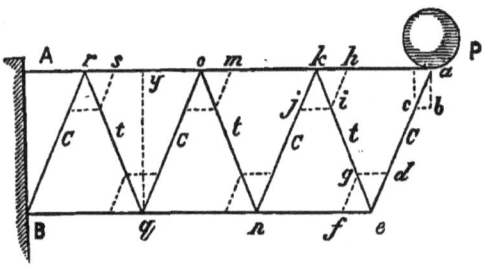

FIG. 56.

until it meets ae in c; then will bc represent the strain on ak, and ac the strain on ae. The weight P is therefore sustained at a by two forces represented by ac and cb. It is evident that ae pushes and ak pulls on P. The force ac is transmitted to e. Take $ed = ac$, and draw dg parallel to en, and prolong it till it meets ek in g. Then will eg represent the strain on ek, and dg that on en. In a similar way, resolve the stresses at k, and so proceed to AB. It will be observed for the

61.—STRAINS ON THE TIES AND BRACES:—1. That the inclined pieces are alternately compressed and extended, as marked in the figure; c indicating compression, and t tension.

2. The amount of compression and tension is the same in all the braces and ties, and equal to $ac = P \sec \theta$. Hence, the vertical component of the strains is uniform over the whole length, and equal to P. This is evidently true whatever be the length of the truss. This is the same as the transverse shearing stress in a beam fixed at one end and free at the other, and has a load P at the free end, $i.\ e.,\ Ss = P$. (See article 3.)

62.—STRAINS ON THE CHORDS.—The pull cb on ak is transmitted to A through the successive parts ko, or, and rA. The pull of ki at k, and the push of kj on the same point, each of which are the same in amount as cb, and which together equal ji, is also transmitted to A, so that the strain of tension on ko is equal to $bc + ji$. In a similar way we find that the strain at A is equal to $bc + hk + mo + sr. = bc + 3\ hk = 7\ bc$. In a similar way we find that the compression on the lower chord at $B = 3\ ef$. We observe:—

1st, That the upper chord is subjected to tension throughout its whole length;

2d, The lower chord is subjected to compression throughout its whole length;

3d, The tension of the upper chord increases from the free end to the fixed end, and that the increments of increase are equal at the successive joints; and

4th, The compression of the lower chord increases by successive equal increments at each of the successive joints, and is a maximum at the fixed end. The increments are the horizontal components of the strains on the ties and braces.

The third and fourth observations are analogous to the moments of stress when a beam is fixed at one end and free at the other, and has a weight P applied at the free end.

Secondly, consider an analytical solution. The strains upon the ties and braces are sufficiently analyzed. To find the strain upon the n-th bay, as ro, of the upper chord, we may at first suppose that this bay is severed, and that a horizontal force t_n equal to the pull on this bay is substituted for the stress. When ro is severed, the truss tends to turn about q. Take q as the origin of moments, and erect qy perpendicular to ro. The lever arm of P is ay; and the lever arm of t_n is qy. Hence, we have

$$t_n \cdot qy = P \cdot ay \therefore t_n = \frac{ay}{qy}P = \frac{2\frac{1}{2}l}{D}P.$$

But $\dfrac{l}{D} = \dfrac{2\ cb}{ab} = \dfrac{2\ cb}{P} \therefore t_n = 5\ cb$. This equals $om + kh + cb = 5\ cb = 5\ P\ tang\ \theta$.

63.—CASE II.—A SEMI-TRUSS HAS EQUAL WEIGHTS AT EVERY JOINT OF THE LOWER CHORD.

Let the truss be like the preceding one inverted, as in Fig.

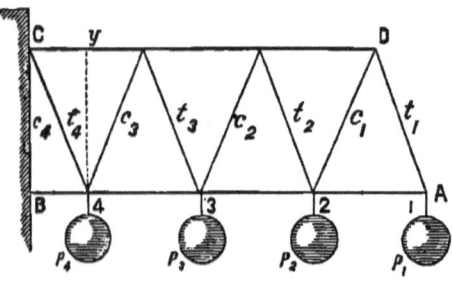

FIG. 57.

57. Let the joint at which the weights are applied be numbered in regular order, beginning with 1 at the free end. For convenience let the weight at 1 be called p_1; that at 2, p_2, and so on.

64.—STRAINS ON THE TIES AND BRACES.—It is evident from the analysis of the preceding case that if p_1 were the only weight on the truss, the vertical component of the strains would be uniform over the whole length and equal p_1; and that the strain on each tie and brace due to this load would be

STRAINS ON A SEMI-TRUSS. 75

p_1 sec θ. Similarly, p_2 would produce a strain of p_2 sec θ, on all the ties and braces between 2 and B. Similarly, p_3 would produce the same strain on all between 3 and B; and so on for any number of weights. Hence, the strain on the first pair of braces (*braces* being used in a general sense to include inclined pieces) will be p sec θ
Stress on the 2d pair $= 2\,p$ sec θ
Stress on the 3d pair $= 3\,p$ sec θ
Stress on the n-th pair $= n\,p$ sec θ..........(98)

We here observe that the strain on the pair of braces between the loaded joints is equal. Also, the first one is subjected to tension, and the next to compression, and so on alternately to the fixed end, as shown in the figure. The wall supports all the weights; hence the vertical force on the wall equals ΣP.

65.—THE STRAINS ON THE CHORDS may be found by adding the components of the strains on the ties and braces as stated in the preceding case. We will thus have for the strain on bay 3—4:—

$c_n = t_1 \sin\theta + c_1 \sin\theta + t_2 \sin\theta + c_2 \sin\theta + t_3 \sin\theta$
$ = p\ tang\ \theta + p\ tang\ \theta + 2\,p\ tang\ \theta + 2\,p\ tang\ \theta + 3\,p\ tang\ \theta$
$ = 9\,p\ tang\ \theta.$

But the more common mode, and one which is generally the easiest, is to find the strain from an equation of moments, thus:— from the joint x directly over 4—3, let fall a perpendicular xz to 4—3. Let the stress on 4—3 be c'. Taking x as the *origin of moments*, and the moment of c' will be $c'xz = c'\,D$.

The moment of $p_1 = p_1 \times zA = 2\tfrac{1}{2}\,lp_1$
The moment of $p_2 = p_2 \times z\,2 = 1\tfrac{1}{2}\,lp_2$
The moment of $p_3 = p_3 \times z\,3 = \tfrac{1}{2}\,lp_3$

Hence, we have, when $p_1 = p_2 = p_3$, etc. $= p$
$c'D = 4\tfrac{1}{2}\,lp$
$\therefore c' = \dfrac{9l}{2D}p = 9\,p\ tang\ \theta$, as before..(99)

Generally, if $n =$ the number of the bay, we have
$c_n = n^2\,p\ tang\ \theta.$

This will give the push on the lower chord at the wall. The tension on the n-th bay of the upper chord is:—

$$t_n = \tfrac{1}{2} n (n + 1) p \frac{l}{D} = n (n + 1) p \, tang \, \theta \ldots (100)$$

If the first bay of the lower chord next to the wall is a full one, as in Fig. 58, the mode of solution is exactly the same, and the strains upon all the parts will be the same as in the preceding case for the same load, except for the piece CE, which is additional in this figure, and will be subjected to a greater strain than any other part of the chords, but the strain is correctly given for this bay by Equation (100).

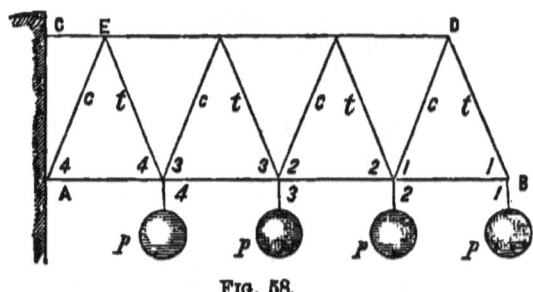

Fig. 58.

66.—OPEN SWING BRIDGE.—When a swing bridge (a kind of draw-bridge) is open, like Fig. 59, it is in the condition of a

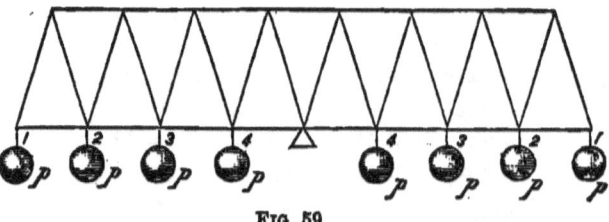

Fig. 59.

girder supported at the middle and uniformly loaded. There may be no live load upon it, and the several weights which are due to the weight of the bridge may, without sensible error, be considered equal on the lower chord, in which case the analy sis of strains is the same as in Case II. If the load is considered upon the upper chord, the analysis is essentially the same.

67.—CASE III.—SUPPOSE THAT A TRIANGULAR TRUSS IS SUPPORTED AT ITS ENDS AND LOADED AT ANY POINT OF THE UPPER CHORD.

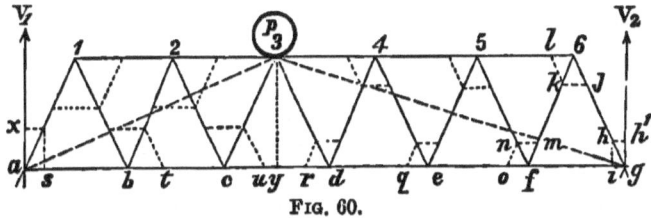

FIG. 60.

Let the truss be represented by Fig. 60, in which the triangles are isosceles. For the present, neglect the weight of the truss.

68.—GEOMETRICAL SOLUTION.—Let the weight p be applied at the third joint in the upper chord. Let V_1 be amount which the support a sustains, and V_2 that sustained at g.

Let fall the perpendicular 3 y. According to the principle of the lever, we have

$$V_1 = \frac{yg}{ag} p \text{ and } V_2 = \frac{ay}{ag} p.$$

Let gh' represent V_2, and construct the parallelogram $gh'hi$, and $hh' = gi$ will be the strain on the chord ga, and $gh =$ the strain on the brace $g6$. gi is transmitted to a through the chord ag; and gh to 6 through the brace $g6$. Take $j6 = gh$ and construct the parallelogram $6jkl$, having $6j$ for one side and jk parallel to the upper chord for the adjacent side. Then $jk = l6$ will be the strain on the first bay of the upper chord, and $k6 = j6 =$ the strain on the tie $f6$. $k6$ will be transferred to f and $fm = k6$, resolved as before. The part fo is transmitted through the chord until it is resisted by an equal strain at the opposite end. Proceed in this way with each of the resultants until we arrive at 3.

Then begin at the opposite end and take $ax = V_1$, resolve it and proceed as before until we arrive at the same point.

We thus find that the strains on the bays of the lower chord are:—on $gf, =$ to ig; on $fe, = gi + fo$; on $ed, = gi + fo + eq$; on $dc, = gi + fo + eq + dr$; and so we might proceed to a, observing to subtract cu, etc., after passing y. But it is better to begin at a and thus find for the strains:—on $ab, = as$; on $bc, = as + bt$; on $cd, = as + bt + cu_1$ which also equals $rd + qe + of + ig$.

69.—OBSERVATIONS ON THE PRECEDING RESULTS.— We observe from this solution that

1st, The strains on all the ties and braces between 3 and g are equal to each other, and that the same is true of those between 3 and a; but the strains on the latter exceed these on the former if $a\,3$ is less than $3\,g$.

2d, The strains on the chords are greatest at and under 3, where the load is applied, and are least at the ends.

We might proceed in this way with weights applied at other points, but a numerical solution shows the results more clearly.

CASE III.—SUPPOSE THAT A TRIANGULAR TRUSS IS LOADED AT EVERY APEX IN THE UPPER CHORD.

Let the truss have equal bays, the triangles isosceles, and let the weights be applied at the joints 1, 2, 3, etc., in the upper chord, as shown in Fig. 61.

70.—DISTRIBUTION OF STRAINS ON THE TIES.—In the following solution of this case, I consider the effect of each weight by itself, and enter in the figure by the side of the piece a number which *indicates* the amount of the strain, and enter it in such a way as to *indicate* the character of the strain, whether it is *plus* or *minus*. This process I call the *distribution of strains*. There may be two subdivisions of this case: First, when the number of bays is even; and second, when it is odd. First, consider the case in which the number of bays in the supported chord is even, as in Fig. 61, where $N = 6$. There will be as many joints in the upper chord as there are bays in the lower. In order to distinguish the several weights from each other, let p_1 be the load at 1; p_2 the load at 2; p_3 the load at 3, and so on. Also, let v_1 be the amount of p_1, which is sus-

TRIANGULAR TRUSS.

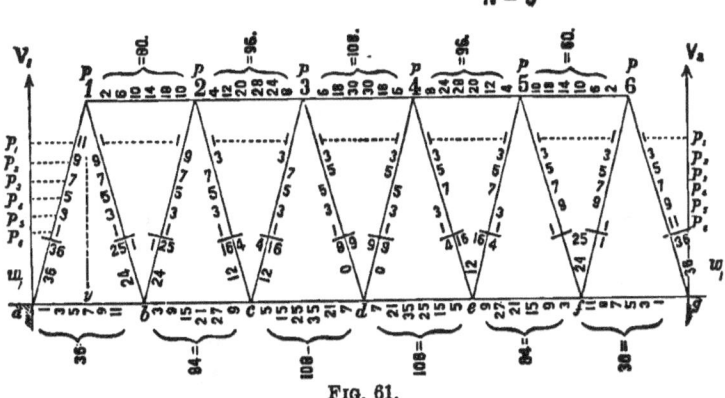

FIG. 61.

tained at a; v_2, the amount of p_2 which is sustained at the same point, and so on, so that $V_1 = v_1 + v_2 + v_3 +$ etc. Similarly, let $V_2 = v' + v'' + v''' +$ etc.

Let fall the perpendicular $1\ y$; then, according to the principle of the lever, we have

$$v_1 = \frac{yg}{ag}p_1 = \tfrac{11}{12} p_1, \text{ and}$$

$$v' = \frac{ay}{ag}p_1 = \tfrac{1}{12} p_1$$

$$\therefore v_1 + v' = p_1, \text{ as it should.}$$

Let $\theta =$ the inclination of the brace to the vertical; then v' resolved in the direction of the brace $g\ 6$ gives for the strain on $g\ 6$ due to p_1; $\tfrac{1}{12} p_1$ sec θ, which is compressive. This is resisted by the strains in $5-6$ and $6f$. The part in $f\ 6$ is tensile and equal to $\tfrac{1}{12} p_1$ sec θ as before, and is transmitted to f; and the same amount by compression to 5; and so on to 1. We observe that all the braces which are inclined towards 1 from the abutment are compressed, and those which incline the opposite way are *tensioned*.*

v_1 resolved in the direction of the brace gives $\tfrac{11}{12} p_1$ sec θ.

---

* This is *coined*, but it comes in here so natural after *compressed*, and is so expressive of the idea which is presented, that we have used it.

The only difference in the value of these expressions is in the numerators, the quantity $\frac{1}{12} p_1 \sec \theta$ being common. Let the numerators represent the relative amount of the strains. These numerators are entered in Fig. 61, opposite p_1, p_2, etc., which are placed at the right and left of the braces in the figure. For compressive strains the numbers are placed on the right-hand side of the tie-brace, and for tensile strains the numbers are placed on the left-hand side. Thus, beginning at the right hand, and opposite p_1, the number 1 is placed on the right side of g 6 ; on the next one it is on the left side ; and so on.

Next consider p_2. We have $v_2 = \frac{9}{12} p_2 \sec \theta$, and $v' = \frac{3}{12} p_2 \sec \theta$. The numbers 9 and 3 are entered as above explained, opposite p_2. Proceed in a similar way with p_3, p_4, and so on to the last.

The figures which are thus entered may be called *coefficients of strains.*

71.—RESULTS.—An examination of the *coefficients of strains* in Fig. 61. readily gives the following results:—

 a. Whether the weights are equal or unequal the end braces are strained most when all the apexes are loaded.

 b. The strains on all the other tie-braces are not a maximum when all the apexes are loaded. For instance, if we consider the brace-tie *c* 3, we observe that p_1 and p_2 produce tension while all the others produce compression, and when all are loaded it is evident that the resultant strain will be the difference of the two strains. If all the weights are equal, the sum of the *coefficients* of compressive strains will be 16, as given in the figure, and the sum of the tensile strains will be 4, which is also given in the figure. The resultant strain will be $- 16 + 4 = - 12$, which is placed on the right-hand side of *c* 3.

 c. Whether the weights are equal or unequal, the strains on the tie-braces between the loaded joints are equal, but in an opposite sense, *i.e.*, if one is compressed the other is *tensioned.*

 d. To produce a maximum strain on any pair, for an uniform load, all the apexes between that pair and the remote end must be loaded, and all the others unloaded. Thus the tie-brace *c* 3 will be compressed an amount represented by 16 if the apexes 3, 4, 5, and 6 are loaded, and 1 and 2 unloaded ; but if either 1 or 2, or both, are loaded, the compressive strain will be diminished by just the amount of tension which they would produce.

e. A maximum, in the opposite sense to the former, may be obtained by loading from the pair considered to the near end and unloading all the rest. Thus, if c 3 has a maximum compression for the loads 3, 4, 5, and 6, it has a maximum *tension* for the loads 1 and 2. Similarly, c 2 has a maximum *tension* for the loads on 3, 4, 5, and 6, but a maximum compression for the loads on 1 and 2. For convenience in the further discussion of similar cases, call the greater maximum, or that given by principle *d*—the **PRIMARY MAXIMUM,** and that given by principle *e* the **SECONDARY MAXIMUM.**

f. Suppose that all the apexes are loaded with equal weights. Add all the *coefficients of strain* for compression, as shown in the figure, and do the same for tension. Take the difference of these and enter them as shown in the figure. We thus find for the resultant strain, 0 for the central pair, 12 for the first pair, each side of the centre, 24 for the next, and so on to the end, from which we observe, in this case, that :—

1st, There is no strain on the central pair of braces ;

2d, The strains upon the tie-braces are proportional to the distances of their lower end from the centre of the lower chord.

Hence, the actual strains on the tie-braces are :—
On the central pair,............................. = 0
On the first pair from the centre, $= 12 \times \tfrac{1}{12} p \sec \theta = p \sec \theta$
On the second pair from the centre, $= 24 \times \tfrac{1}{12} p \sec \theta = 2 p \sec \theta$
On the third pair from the centre, $= 36 \times \tfrac{1}{12} p \sec \theta = 3 p \sec \theta$
On the x pair from the centre, $= x p \sec \theta$ (101).

With the exception of $\sec \theta$, these results are the same as for the shearing stress on a horizontal beam which is loaded with equal weights at equi-distant points, and symmetrically placed in reference to the centre of the beam. See Eq. (17.)

g. When the weights are unequal, any joint may be loaded so heavily as to determine the *nature of the strain* on all the tie-braces, in which case all those which incline towards the load will be compressed, and those which incline in the opposite direction will be extended.

If all the weights are equal, and the central pair, p_3 and p_4 be removed, there will be no strain on all the tie-braces between 2 and 5. Similarly, if p_2 and p_5 also be removed, there will be no strain on all between 1 and 6.

72.—FORMULAS FOR STRAINS ON THE TIE-BRACES FOR A UNIFORM LOAD.

Let N = the total number of bays in the supported chord, also equal to the number of joints in the unsupported chord;

n = the number of bays from one end to the foot of the pair which is considered;

x = the number of bays from the centre to the same pair, and

$$p = p_1 = p_2 = p_3 = p_4, \text{ etc.}$$

We have

$$n + x = \tfrac{1}{2} N, \text{ when } n < \tfrac{1}{2} N, \text{ and}$$
$$n - x = \tfrac{1}{2} N, \text{ when } n > \tfrac{1}{2} N$$
$$\therefore x = \pm (\tfrac{1}{2} N - n),$$

which in Eq. (101) gives

$$xp \sec \theta = \pm (\tfrac{1}{2} N - n) p \sec \theta = \mp (N^2 - 2Nn) \frac{p}{2N} \sec \theta$$

for the strain on the tie-braces which *terminate at the end of the n-th bay*. In the last form the quantity in the parenthesis is the *coefficient of strains*.

This equation may also be deduced from the principle of shearing stress. For we have $V_1 = \tfrac{1}{2} Np$, and between the end and n-th bay the load is np; hence, according to Eq. (17), we have $S_n = \tfrac{1}{2} Np - np = (\tfrac{1}{2} N - n) p = \tfrac{1}{2} (N - 2n) p$, and this resolved in the direction of the brace gives

$$\tfrac{1}{2} (N - 2n) p \sec \theta = (\tfrac{1}{2} \Sigma p - n p) \sec \theta \dots (102)$$

In this case we observe that the negative values, which result for $n > \tfrac{1}{2} N$, apply beyond the centre.*

---

* I have sought for a formula which will give the strains on the ties and braces by the simple substitution of the successive numbers 1, 2, 3, etc., the results of which would give the strains on the successive ties and braces

TRIANGULAR TRUSS. 83

For the end braces, a 1 or g 6, make $n = 0$ or $= N$, and Eq. (102) gives

$$+ \tfrac{1}{2} Np \sec \theta = \tfrac{1}{2} W_t \sec \theta \dots \dots \dots (103)$$

in which W_t is the total weight of the load.

This solution suggests another mode of considering the problem. Thus, in Fig. 61, the stress on the end braces is $\tfrac{1}{2} \Sigma p \sec \theta$. To find it on the next pair, conceive that the truss is sup-

counting from one end, as in Fig. 61-a. This I have done by trial. We must be guided by Equation (102), and must make such an equation as that the results shall be the same when we substitute 2, as when we substitute 3; similarly, they must be the same for 4 and 5, and so on. This I do by making such an expression as that one term shall disappear when the other is real. The *signs* of the results must also be *minus* for the odd numbers from the end to the centre, and *plus* beyond the centre;—and *plus* for the even numbers between the end and centre. These results may doubtless be secured in various ways, but I have hit upon the following :—

First, let n' = an odd number of tie-braces = 1, 3, 5, etc., then $n' = 2n + 1$; and Eq. 102 becomes $\tfrac{1}{2}\left[N - n' + 1\right] p \sec \theta$. This term must be so affected as to be minus for the first part of the truss, and at the same time reduce to zero when an even number is substituted for n'.

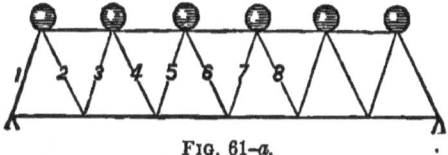

FIG. 61-a.

We observe that $\cos 2 n'\pi$ is always $+1$ for integral values of n'; and evidently $\cos 2 n'\pi - \cos 2 n'\pi = 0$. The odd powers of (-1) are -1, and the even powers are $+1$. $\therefore (-1)^{n'} \cos 2 n'\pi$ is -1 for all odd values of n'. Hence

$$\tfrac{1}{2}\left[(-1)^{n'} \cos 2 n'\pi - \cos 2 n'\pi\right]$$

is -1 for the odd values of n, and zero for the even values. Thus,

If $n' = 1$, we have $\tfrac{1}{2}\left[(-1) \times 1 - 1\right] = -1$.

If $n' = 2$, we have $\tfrac{1}{2}\left[(+1) \times 1 - 1\right] = 0$.

If $n' = 3$, we have $\tfrac{1}{2}\left[(-1) \times 1 - 1\right] = -1$,

and so on. Hence, for the odd numbers of the tie-braces, we have

$$\tfrac{1}{2}\left[N - n' + 1\right] p \sec \theta \left[(-1)^{n'} \cos 2 n'\pi - \cos 2 n'\pi\right]$$

ported at 1 and 6, the points where the first weights are applied counting from each end. Then the stress on 1 b and b 2 will be the half sum of the remaining weights multiplied by $\sec \theta$; which is the same as $(\frac{1}{2}\Sigma p - 1 p) \sec \theta$. To find it on the next pair, conceive that the truss is supported at 2 and 5, and we have the stress on 2 c and c 3 equal to one-half the sum of the remaining weights multiplied by $\sec \theta$ or $(\frac{1}{2}\Sigma p - 2p) \sec \theta$. Hence, generally we have $(\frac{1}{2}\Sigma p - np) \sec \theta$ as before.

If $n' =$ an even number of the tie-braces, we have $n' = 2 n$, and Eq. (102) becomes $\frac{1}{2} (N-n')$, which expression must be positive for even values of n', and disappear for odd values; and as it is to be real when n' disappears, we will use n' instead of n''. Reasoning in exactly the same way as before, and we have for this term

$$\tfrac{1}{2}\left[N-n'\right] p \sec \theta \left[(-1)^{n'} \cos 2\, n'\pi + \cos 2\, n'\pi\right]$$

Hence, *we have for the strain upon the* n^t-*th tie-brace.*

$$\left\{\left[N-n'+1\right] \times \left[(-1)^{n'} \cos 2\, n'\pi - \cos 2\, n'\pi\right] + \left[N-n'\right] \times \right.$$

$$\left. \left[(-1)^{n'} \cos 2\, n'\pi + \cos 2\, n'\pi\right]\right\} \tfrac{1}{4} p \sec \theta \ldots\ldots\ldots\ldots(103a)$$

EXAMPLE.—Let $N = 6$ as in Fig. 61.

Then for the first brace $n' = 1$ and we have $\left[6 \times (-2) + 5 \times 0\right] \frac{1}{4} p \sec \theta = -3 p \sec \theta$

Then for the second tie-brace $n' = 2$ and we have $\left[5 \times (0) + 4 \times (+2)\right] \frac{1}{4} p \sec \theta = +2 p \sec \theta$

Then for the third tie-brace $n' = 3$ and we have $\left[4 \times (-2) + 3 \times (0)\right] \frac{1}{4} p \sec \theta = -2 p \sec \theta$

Then for the fourth tie-brace $n = 4$ and we have $\left[3 \times (0) + 2 \times (+2)\right] \frac{1}{4} p \sec \theta = +p \sec \theta$

Then for the fifth tie-brace $n' = 5$ and we have $\left[2 \times (-2) + 1 \times (0)\right] \frac{1}{4} p \sec \theta = -p \sec \theta$

Then for the sixth tie-brace $n' = 6$ and we have $\left[1 \times (0) + 0 \times (+2)\right] \frac{1}{4} p \sec \theta = 0$

Then for the seventh tie-brace $n' = 7$ and we have $\left[0 \times (-2) - 1 \times (0)\right] \frac{1}{4} p \sec \theta = 0$

Then for the eighth tie-brace $n' = 8$ and we have $\left[-1 \times (0) - 2 \times (+2)\right] \frac{1}{4} p \sec \theta = -p \sec \theta$

Then for the ninth tie-brace $n' = 9$ and we have $\left[-2 \times (-2) - 3 \times (0)\right] \frac{1}{4} p \sec \theta = \times p \sec \theta$

and so on. It will be observed that the signs change after passing the middle, as they should. The signs take care of themselves, and the result tells whether a piece is a brace or tie.

The trigonometrical term may properly be called a *Modulus of Signs.*

TRIANGULAR TRUSS.

73.—ANOTHER SOLUTION. The strain on the end braces is $[1 + 3 + 5 + 7 + \text{etc., to } N \text{ terms}] \frac{p \sec \theta}{2 N} = \frac{1}{2} N p \sec \theta$ as before. The strain upon the pair at the end of the n-th bay is $\left[(1 + 3 + 5 + 7 \ldots \text{ to } (N - n) \text{ terms}) - (1 + 3 + 5 + \ldots \text{ to } n \text{ terms})\right] \frac{p \sec \theta}{2 N}$

$$= \left[(N - n)^2 \frac{p \sec \theta}{2N} - n^2 \frac{p \sec \theta}{2N}\right] = \left[N^2 - 2nN\right] \frac{p \sec \theta}{2N} = \frac{1}{2}\left[N - 2n\right] p \sec \theta$$

as before.

74.—MAXIMUM STRAINS ON THE TIE-BRACES.—The conditions for a maximum strain on the tie-braces have been given in d, Article 71. From the figure we see that the *primary* maximum strain on a pair at the end of the n-th bay is

$$\left[1 + 3 + 5 + \ldots \text{ to } (N-n) \text{ terms}\right] \frac{p \sec \theta}{2N} = (N-n)^2 \frac{P \sec \theta}{2N}$$

This may also be found by the principle of shearing stress; for the load is $(N-n)p$ to produce a maximum strain, and from the principle of the lever, we have

$$V_1 N l = (N-n) p \times \tfrac{1}{2} (N-n) l;$$

$$\therefore V_1 = (N-n)^2 \frac{p}{2N} ;$$

and as there is no load between V_1 and the pair considered, this equals the shearing stress. This multiplied by $\sec \theta$, gives the actual strain; or *maximum strain on the braces at the end of the n-th bay is*

$$(N-n)^2 \frac{p \sec \theta}{2 N} \ldots \ldots \ldots \ldots \ldots (104)$$

If the moments be taken about the other end, we have

$$V_2 N l = (N-n) p \times \left[N - \tfrac{1}{2}(N-n)\right] l$$

$$\therefore V_2 = (N - n)(N + n) \frac{p}{2N}$$

We also have $V_1 + V_2 = (N - n) p \therefore V_2 = (N - n) p - V_1$.
But $S_2 = V_2 - (N - n) p = - V_1$ as before, only with a contrary sign.

Secondary maximum. An examination of the case shows that we may still consider $(N - n)$ joints loaded, but n must exceed $\frac{1}{2}N$. Hence Eq. (104) gives the strain for the *secondary maximum* when $n > \frac{1}{2}N$.*

75.—STRAINS ON THE CHORDS.—*Distribution of the strains.*

First consider p_1. $\frac{1}{12} p_1$ is supported at g and $\frac{11}{12} p_1$ at a, as before stated. The former is to be resolved in the direction of the brace $g6$ and of the chord gf. That in the direction of $g6$ has already been considered. That in the direction of the chord is evidently $\frac{1}{12} p_1$ $tang\ \theta$. Taking 1 to represent the strain as before, and it is placed on the lower side of gf, to represent tension. This strain is transmitted through the successive bays until it is met by a counter-strain from the opposite end. At f, the tie $f6$ produces a pull of $\frac{1}{12} p_1$ $tang\ \theta$, and the brace $f5$ a push of the same amount, so that both produce an increment of $\frac{2}{12} p_1$ $tang\ \theta$; which added to the $\frac{1}{12} p_1$ $tang\ \theta$ previously found gives $\frac{3}{12} p_1$ $tang\ \theta$ for the total strain on fe. Similarly, we find $\frac{5}{12} p_1$ $tang\ \theta$ for the strain on ed, and so on to ba, where it is $\frac{11}{12} p_1$ $tang\ \theta$. Proceeding in this way with each of the weights, observing to pass from each end to the bay directly under the weight, and we readily find all the numbers which are placed below the lower chord. In the same way we find those which are placed over the upper chord.

We observe from the figure

1st, That the strains are all in the same sense, *i. e.*, all the weights produce tension on the lower chord, and compression on the upper; and hence,

2d, The chords receive their maximum strain when all the joints are loaded; and

3d, If all the weights are equal the greatest strain will be at the middle of the chords and least at the ends. This is similar

---

* Proceeding as in the preceding *foot-note*, and we may find for the maximum strain on the n'-th tie-brace the following expression:—

$$= \left[2\ N - n' - 1\right]^2 \times \left[(-1)^{n'} \cos 2n'\pi - \cos 2n'\pi\right]$$
$$+ \left[2\ N - n'\right]^2 \times \left[(-1)^{n'} \cos 2n'\pi + \cos 2n'\pi\right] \frac{p \sec \theta}{16\ N}$$

TRIANGULAR TRUSS. 87

to a beam which is supported at its ends and uniformly loaded over its whole length.

This mode of determining the strains on the chords is not considered practical.

The method by moments is the usual one.

From 1 let fall the perpendicular $y1$

Let $y1 = D =$ the depth of the truss;
  $t_1 =$ the tension on ab, the first bay;
  $t_2 =$ the tension on bc, the second bay;
  $t_n =$ the tension on the n-th bay; and
  $l = ab =$ the length of a bay.

Take the origin of moments at 1, the point about which the truss will turn if ab be severed. When the internal forces are substituted for the strains, and we consider the conditions of equilibrium in reference to them, we must consider the external forces only on one side of the section; for the internal forces transmit the strains from one side of the section to the other. Hence, for equilibrium, we have

$$V_1 \, ay = t_1 \, y1$$

$$\therefore t_1 = \frac{ay}{y1} V_1 = \frac{\frac{1}{2}l}{D} V_1$$

for the tension on the first bay.

For the second bay take the origin of moments at 2, the point about which the frame would turn in case bc were severed. Hence we have

$$V_1 \times \tfrac{3}{2} l - p_1 \times l = t_2 \times D$$

$$\therefore t_2 = (3 \, V_1 - 2 p_1) \frac{l}{2 D},$$

for the tension on the second bay. Similarly,

$$t_3 = (5 V_1 - 4 p_1 - 2 p_2) \frac{l}{2 D},$$

$$t_4 = (7 V_1 - 6 p_1 - 4 p_2 - 2 p_3) \frac{l}{2 D}$$

$$t_n = \Big((2n-1) V_1 - (2n-2) p_1 - (2n-4) p_2 \dots p_n\Big) \frac{l}{2 D}. \quad (105)$$

If $p = p_1 = p_2 = p_3 =$ etc. $= p_n$ the expression becomes much simplified, but instead of reducing the preceding equation for this case, we may solve it directly from the figure, thus

$$V_1 = \tfrac{1}{2} N p$$

Moment of $V_1 = V_1 (n - \tfrac{1}{2}) l = \tfrac{1}{4} N p (2n - 1) l$, which is the moment of an upward force. There will be n downward loads including the one over the n-th joint;* hence the downward load will be $n p$. The abscissa of the centre of gravity of this load from the origin of moments, is $\tfrac{1}{2} (n - 1) l$; and hence the moment of this load is $\tfrac{1}{2} n (n - 1) p l$. The moment of the tension is $t_n D$. Hence we have for

The tension on the n-th bay of the lower chord

$$t_n = \left[N(2n - 1) - 2n(n - 1) \right] \frac{p l}{4 D} \ldots \ldots (106)$$

EXAMPLE.—Let $N = 6$; $\dfrac{\tfrac{1}{4} l}{D} = tang\ \theta$. Then Equation (106) gives

For $n = 1,\ t_1 = \tfrac{11}{1} p\ tang\ \theta$
$n = 2,\ t_2 = \tfrac{21}{1} p\ tang\ \theta$
$n = 3,\ t_3 = \tfrac{108}{12} p\ tang\ \theta$
$n = 4,\ t_4 = \tfrac{108}{12} p\ tang\ \theta$
$n = 5,\ t_5 = \tfrac{21}{1} p\ tang\ \theta$
$n = 6,\ t_6 = \tfrac{11}{1} p\ tang\ \theta$

It will be observed that the numerators, 36, 84, 108, &c., are the sums of the strains as given in Fig. 61.

In Eq. (106), if $n = \tfrac{1}{2} N$, $t_n = \tfrac{1}{8} \dfrac{N^2 p l}{D} = \tfrac{1}{8} \dfrac{W_s L}{D}, \ldots (107)$

in which

$W_s =$ the weight of the total surcharge,
$L =$ the span, and
$D =$ the depth.

Strains on the upper chord.

Let $c_n =$ the compression on the n-th bay of the upper chord, and the other notation as before.

Let all the weights be equal to each other. If any bay of the upper chord be severed, the system will turn about some joint of the lower chord. Take the origin of moments at

---

* The moment of the load at the joint about which it tends to turn is zero; hence it will make no difference whether we include this or not. For the lever arm will be different, and the result will be the same in both cases. If it is not included, the load will be $(n - 1) p$.

any joint (the n-th) of the lower chord. The moment of V_1 will be $V_1 nl = \frac{1}{2} Npnl$. There will be a load equal to np acting down, whose lever arm (or distance from the origin of moments to the centre of the loading) will be $\frac{1}{2}nl$; hence the moment of the downward forces will be $\frac{1}{2}n^2pl$. Hence

The strain on the n-th bay of the upper chord will be

$$c_n = \left(\tfrac{1}{2} Npnl - \tfrac{1}{2}n^2 pl \right) \frac{1}{D} = (N-n)\frac{npl}{2D} \ldots \ldots (108)$$

that is, the strains vary as the product of segments into which the chord is divided by the joint considered.

If $n = \tfrac{1}{2} N$ we have

$$c_n = \tfrac{1}{8} N^2 \frac{lp}{D} = \tfrac{1}{8} W, \frac{L}{D}$$

EXAMPLE.—Let $N = 6$, and $\dfrac{\tfrac{1}{2} l}{D} = $ tang θ. Then Eq. (108) gives,

For $n = 1$, $c_1 = \tfrac{60}{12} p$ tang θ
$n = 2$, $c_2 = \tfrac{96}{12} p$ tang θ
$n = 3$, $c_3 = \tfrac{108}{12} p$ tang θ, etc.

The numerators of which are the same as the sum of the strain, in the several bays of the upper chord in Fig. (61).

76.—SUB-CASE, IN WHICH THE NUMBER OF BAYS IN THE LOWER CHORD IS ODD.—In the preceding case the number of bays in the supported chord was even. Now suppose that they are odd, as in Fig. 62, where $N = 5$.

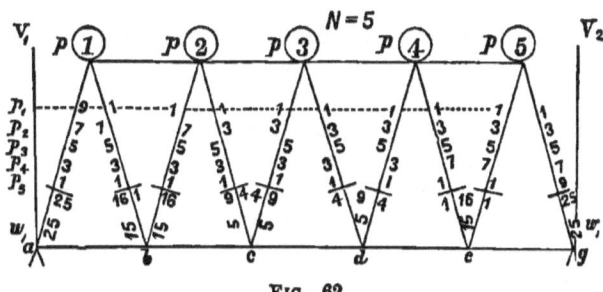

FIG. 62.

Distributing the strains in exactly the same way as before, and we have the numbers as given in the figure. A comparison of Fig. 62 with Fig. 61 shows that all the principles which were drawn from Fig. 61 are applicable to this, except that of the strains on the central braces. There it was zero,—here it is $\frac{5}{10} p \sec \theta = \frac{1}{2} p \sec \theta$.

The cause of this difference is due to the fact that in this case there is a load p at the middle joint, which was not at that point, in the former case. The strain on the central pair here is the same as if no load was upon the truss except the one at the third joint.

The *formulas* for the strains upon the chords and tie-braces are the same as in the preceding case. Hence Eqs. (102) (104) (106) and (108) are directly applicable.

77.—INVERTED TRUSS.—If the truss is supported by the upper chord, as in Fig. 63, the *nature* of the strains will all be inverted; that is, the supported chord will be *compressed* instead of *tensioned*, and the end tie-braces become *ties* instead of *braces*. But the amount of the strains on all the pieces will be the same as in the preceding case.*

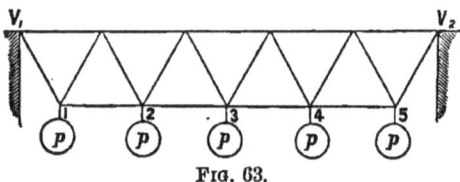

Fig. 63.

78.—WEIGHT OF THE TRUSS CONSIDERED.—Thus far no account of the weight of the truss has been considered. This is a *permanent* dead load. Its amount cannot be accurately determined until the dimensions have been determined, and these cannot be ascertained until the *total* load is known. Hence, algebraically speaking, the weight of the truss is an implicit function of its dimensions, and in most cases it is impracticable if not impossible to make it explicit. The practical way is to *assume* the weight of the truss and find its dimensions on this hypothesis; then compute the weight from these dimensions

---

* To make Eq. (103 *a*) strictly applicable to this case will require a *minus* sign before the whole expression.

TRIANGULAR TRUSS. 91

and compare the result with the assumed weight. If the bridge is properly proportioned, we see from the preceding analysis that the chords will be largest at the middle and smallest at the ends; and the braces will be largest at the ends and smallest at the middle. We may, therefore, without sensible error assume that the weight of the bridge is a permanent uniform load. If we further assume that the weight of the bridge is carried at the upper apexes, we have

$$\frac{W_1}{N} = w_1 = w_2 = w_3 = \text{etc}\ldots\ldots\ldots(109)$$

in which

$W_1 =$ the weight of the truss;
$N =$ the number of bays in the lower chord; and
$w_1 =$ the weight on each joint due to the weight of the truss.

This value of w_1 substituted in Eq. (102) gives

For the strain on the tie-braces at the end of the n-th bay due to the weight of the truss,

$$= \tfrac{1}{2}(N - 2n)\, w_1 \sec\theta \ldots\ldots(110)$$

The differences 36, 24, 12, etc., placed opposite the tie-braces in Fig. 61 may be considered as coefficients of $\dfrac{w_1}{2N}\sec\theta$.

Adding Equations (104) and (110), and we have

The strains on the tie-braces at the end of the n-th bay when the weight of the truss is considered, and a uniform load extends from the n-th bay to the opposite end, both loads on the upper chord;

$$= \left[(N-n)^2 p + (N-2n)\, N w_1\right]\frac{\sec\theta}{2N} *.(111)$$

---

* Introducing the moduli of signs and we have the strain on the n'-th tie-brace

$$= \left\{\left[(2N - n' - 1)^2 p + 4(N - n' + 1)\, N w_1\right] \times \left[(-1)^{n'}\cos 2n'\pi - \cos 2n'\pi\right] + \left[(2N - n')^2 p + 4(N - n')\, N w_1\right] \times \left[(-1)^{n'}\cos 2n'\pi + \cos 2n'\pi\right]\right\}\frac{\sec\theta}{16}$$

79.—DISCUSSION OF EQUATION (111).—1st, If $n = \frac{1}{2} N$, the term containing w_1 disappears; that is, when the number of bays in the supported chord is even, the truss as an uniform load produces no strain on the central *tie-braces;* and the Equation becomes

$$\tfrac{1}{2} N p \sec \theta = \tfrac{1}{2} W_1 \sec \theta \ldots\ldots\ldots\ldots (112\ a)$$

for the maximum strain on the central tie-braces.

2d, If $n = 0$, we have

$$\tfrac{1}{2}(Np + Nw_1) \sec \theta = \tfrac{1}{2} W \sec \theta \ldots\ldots (113\ b)$$

which gives the strain on the end braces. It always exceeds four times the maximum strain on the central pair.

3d, If $N = 1$, we have the case of a pair of rafters, and at the same time making $n = 0$, and we have for the strain on each rafter $\tfrac{1}{2}(p + w_1) \sec \theta$, which is the same as the second of Eqs. (68).

4th, If $n < \tfrac{1}{2} N$, the coefficients of p and w_1 are both positive, and hence the weight of the bridge and live load *conspire* to produce strains when the live load extends over more than half the length of the truss.

5th, If $n > \tfrac{1}{2} N$, the coefficient of w_1 becomes *negative*, and the other term remains *positive*, and hence the dead and live loads act in opposite senses, and hence the total strain is the difference of the two.

6th, If Eq. (111) be placed equal zero, and solved for n, we have

$$n = \left[\frac{p + w_1}{p} \pm \sqrt{\left(\frac{p + w_1}{p}\right)^2 - \frac{p + w_1}{p}} \right] N \ldots (112)$$

It may easily be shown that there will always be one value of n in Eq. (112), which is less than N and greater than $\tfrac{1}{2} N$. Call this value n_o (for it reduces Eq. (111) to zero). The other value of n exceeds N, and hence is beyond the limits of the problem. At n_o the vertical shearing stress is zero.

In order to get a better idea of the results of this discussion, conceive in Fig. 61 that all the apexes of the upper chord are loaded with equal weights, then the end brace will receive its maximum strain, which is found by making $n = 0$ in Eq. (111). Now remove p_1 and the tie $b\ 1$ and brace $b\ 2$ will receive their maximum strain, and may be found by making $n = 1$ in Eq. (111). Remove p, and the tie-braces $c\,2$ and $c\,3$ receive their maximum strain, and so on, to the middle.

As soon as we pass the middle, the resulting strains will be in the contrary sense to that which is produced by the permanent load, and there will be a *secondary* maximum, the *primary* maximum being secured by the load passing off in the opposite direction. Continuing to remove the weights beyond the centre, we soon pass the point n_o, and after that *the only effect of the remaining weights as they are removed one after the other will be to relieve the strains which are produced by the weight of the bridge*. Equation (111) for $n > n_o$ becomes negative. By moving in the opposite direction, the distance n_o will be counted from g. It will be observed that n represents the unloaded part. The *tie-braces* each way from the middle to the joint n_o are true *tie-braces*, for they will be subjected to both tension and compression for loads which move on or off either way. And generally we observe that in any case those inclined $\left\{\begin{array}{c}\text{towards}\\\text{from}\end{array}\right\}$ the point of zero shearing stress will act as $\left\{\begin{array}{c}\text{braces}\\\text{ties.}\end{array}\right\}$

EXAMPLE.—Let $p = 6\ w_1$ and $N = 6$

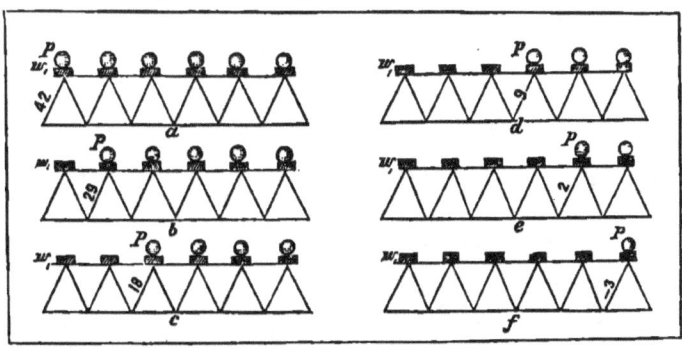

FIG. 63–a.

Let the rectangles on the upper chords represent w_1, and the spheres p. In Equation (111) make $N = 6$ and $p = 6\ w_1$ and it becomes $\left[(6-n)^2 + 2(3 - n)\right] p_1 \frac{\sec\theta}{12}$ When $n = 0$, we have the case shown by (a) Fig. 63–a, and the strain is, $\tfrac{4\,2}{1\,2} p_1 \sec\theta$.

Retaining only the quantity in the parenthesis and we have for the maximum strain on all the tie-braces for a uniform load moving off without shock, by making $n = 0, 1, 2, 3$, etc., in Eq. (111), as follows:—

For $n = 0$, strain on end brace or 1st half pair, $= 42$, *see a* Fig.63-*a*
$n = 1$, strain on end brace or 2d pair, $= 29$, *see b* Fig.63-*a*.
$n = 2$, strain on end brace or 3d pair, $= 18$, *see c* Fig.63-*a*.
$n = 3$, strain on end brace or 4th pair, $= 9$, *see d* Fig.63-*a*.
$n = 4$, strain on end brace or 5th pair, $= 2$, *see e* Fig.63-*a*.
$n = 5$, strain on end brace or 6th pair, $= -3$, *see f* Fig.63-*a*.
$n = 6$, strain on end brace or 7th half pair $= -6$,

Those which incline towards the rear end of the load act as braces, with the exception of the second one from the end which remains a tie. The last one is the strain on the end brace due to the weight of the truss.

80.—PROBLEM.—Find the Strain upon *any* tie-brace—the x-th for instance—when the n-th is strained to a maximum. This is equivalent to finding the strain upon any tie-brace for any of the cases shown in Fig. 63-*a*.

81.— STRAINS ON THE CHORDS WHEN THE WEIGHT OF THE TRUSS IS CONSIDERED.

Let $W = W_1 + W_2 =$ the total load; then $p + w_1 = W \div N$, which may be substituted directly in Equations (106) and (108). Hence, we have for the strain on the n-th bay of the lower chord

$$t_n = \left[N(2n - 1) - 2n(n - 1) \right] \frac{p + w_1}{4D} l \ldots \ldots (113)$$

And for the compression on the n-th bay of the upper chord

$$c_n = (N - n) \frac{nl}{2D} (p + w_1) \ldots \ldots \ldots \ldots (114)$$

82.—DIMENSIONS OF THE TIES.

Let $K =$ the section of a tie;
 $T =$ the modulus of tenacity; and
 $f =$ the factor of safety $= \frac{1}{4}$ to $\frac{1}{5}$ for iron, and $\frac{1}{10}$ for wood.

Then $fKT =$ the working resistance, which must equal the strain given by Eq. (111);

$$\therefore fKT = \left[(N - n)^2 p + (N - 2n) Nw_1 \right] \frac{\sec \theta}{2N}$$

$$\therefore K = \left[(N-n)^2 p + (N-2n) Nw_1 \right] \frac{\sec \theta}{2fTN} \ldots (115)$$

In which for a maximum n should not exceed $\frac{1}{2} N$. If several ties are used to resist the strain at any point, this value must be divided between them. In the case of bridges there are always two or more trusses, and the above values should be divided by two or more, and sometimes there are two or more ties in the same truss to resist the strains at any point. When n exceeds $\frac{1}{2} N$ in Eq. (111), the resulting values of the equation show the strains which will fall upon the fastenings at the end when a piece which is ordinarily a brace becomes a tie. In wooden structures the parts are rarely proportioned so as to vary according to the strains, but they are proportioned to carry the greatest strains, and all the other pieces are made of the same size. Conventional sizes are also sought even at the expense of some waste of timber, for the aggregate cost is generally diminished by so doing.

83.—DIMENSIONS OF THE BRACES.—The braces are generally so long that they may be considered as short or else as long columns.

Let b = one side of a brace if it is rectangular;
d = the other, $b < d$;
m = the number of braces used to carry a strain;
f = the factor of safety; and
Then from (58), we have

$$b^2 = \left[(N-n)^2 p + (N-2n) Nw_1 \right] \frac{6 l^2 \sec \theta}{\pi^2 Emfd N}$$

If the braces are round, their total section will be $m\pi r^2$ in which r is the radius of the cylindrical column; and r may be found from Equations (60), (61), (62), (63), (64), or (65), as the case may be, combined with Equation (111).

84.—DIMENSIONS OF THE CHORDS.—The dimensions of the chords may be found in the same way as the ties and braces.

CASE IV.—TRIANGULAR TRUSS LOADED AT THE JOINTS OF THE SUPPORTED CHORD.

85.—DISTRIBUTION OF STRAINS.—In Fig. 64 the number of bays in the supported chord is odd, and equal to five. As

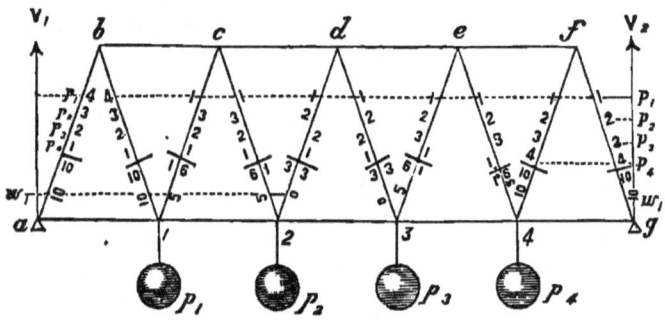

FIG. 64.

before, calling the weight at 1, p_1, at 2, p_2, etc., and we find that V_1 sustains four-fifths of p_1; and V_2, one-fifth of p_1. The proper figures are entered on the upper part of the truss, those representing compression being on the right, and tension on the left of the respective pieces, as in the preceding case.

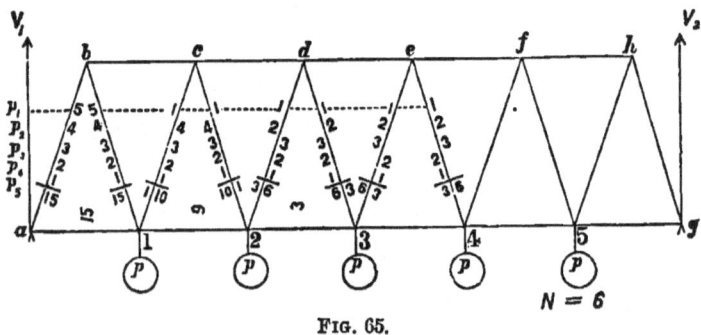

FIG. 65.

In a similar way we find the distribution of the strains when the number of bays is even (six) as in Fig. 65.

An examination of Figs. 64 and 65, compared with Articles 97 and 102, shows that they present no new *principles*. There is a difference in the numerical results which is due to the fact that the load in this case is supported on one less apex than there are bays. We observe here that the two tie-braces at each end receive the same strain, so that we may count them by pairs, the n-th pair corresponding to the n-th bay.

86.—STRESS ON THE TIE-BRACES.—1st, *For equal weights on every joint.*

The total load $= (N-1) p$

$$\therefore V_1 = V_2 = \tfrac{1}{2} (N-1) p$$

The number of weights between the end, as a, and the n-th pair, is $n-1$. Hence, the shearing stress on the n-th pair is

$$S_t = V_1 - (n-1) p = \tfrac{1}{2}(N-1) p - (n-1) p = \tfrac{1}{2}(N-2n+1) p$$

$\therefore$ strain on n-th pair due to the p's is,

$$= \tfrac{1}{2} (N -2 n +1) \, p \sec \theta \dots\dots\dots\dots\dots\dots\dots\dots\dots (116)$$

strain on n-th pair due to the weight of truss is

$$= \tfrac{1}{2} (N -2 n +1) \, w_1 \sec \theta \dots\dots\dots\dots\dots\dots\dots\dots (117)$$

and the strain on n-th pair due to the p's, and weight of truss, is

$$= \tfrac{1}{2} (N - 2 n + 1) \, (p + w_1) \sec \theta \dots\dots\dots\dots (118)$$

In this case the total weight of the truss is supposed to be supported at the joints of the lower chord. This is done for convenience in the investigation. A part of the weight is evidently supported at the joints in the upper chord, and the remainder at the joints of the lower chord. This modification of the case will be considered hereafter.

The above result may also be found by summing the series of coefficients, and multiplying the result by $p \sec \theta$, as in the preceding case. Thus,

$$\left[1+2+3+4+ \; to \; (N - n) \; terms \; -(1+2+3+\dots\dots \dots to \, (n-1) \; terms) \right] \frac{p \sec \theta}{N} = \tfrac{1}{2} N(N-2n+1) \frac{p \sec \theta}{N}$$

2d, *Maximum strain for equal weights.* An examination of Figs. 64 and 65 shows the same result as before, *that to produce a maximum stress on the n-th pair of braces the* *load must extend from that pair to the remote end, and the remaining part be unloaded.* This principle is general for all trusses with parallel horizontal chords. In this case we have $(n-1)$ unloaded joints, and $N-n$ loaded ones. Hence, for the maximum stress on the n-th pair of braces we have

$$\left[1+2+3+ \text{etc.}, \ldots \text{to } (N-n) \text{ terms}\right]\frac{p \sec \theta}{N} = (N-n)$$

$$(N-n+1)\frac{p \sec \theta}{2N} \ldots \ldots \ldots \ldots \ldots \ldots \ldots \ldots (119)$$

Or by the principle of shearing stress, we have

$$S_s = V_1 = (N-n)(N-n+1)\frac{p}{2N}$$

or

$$S_s = V_s - (N-n)p = \left[(N-n)p - V_1\right] - (N-n)p = -V_1$$

as before, with a contrary sign. This multiplied by $\sec \theta$ gives the stress as before. In the former expression, $S_s = V_1$ because there is no load between V_1 and the n-th pair to be subtracted.

3d, *Maximum stress for a partial uniform live load and a total dead load.* The latter includes the weight of the truss. Adding Equations (117) and (119) gives

The maximum strain on the n-th pair of tie-braces for uniform weights including the weight of the truss.

$$= \left[(N-n)(N-n+1)p + (N-2n+1)Nw_1\right]\frac{\sec \theta}{2N} \quad (120)$$

in which for a primary maximum $n = $ or $< \frac{1}{2}N$.

This equation may be discussed in the same way as Eq. (111).*

---

* By numbering the tie-braces as in Fig. 66, and introducing *a modulus of signs*, we have

The maximum strain on the n'-th tie-brace.

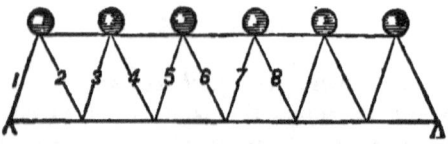

Fig. 66.

87.—GENERAL METHOD.—For parallel horizontal chords the shearing stress multiplied by the secant of the inclination of the tie-braces, gives the stress on the ties and braces at that section.

The reason is evident, for the chords do not sustain any transverse stress (when the joints are perfectly flexible as we are now supposing), and hence it is all sustained by the ties and braces. The value of the transverse shearing stress is given in Article 84 of the author's "Resistance of Materials." It is simply the resultant vertical force. Hence, if

V_1 = the reaction of one support;
V_2 = the reaction of the other support;
$\Sigma_0^n p$ = the sum of all the weights between V_1 and the end of the n-th bay; and
$\Sigma_0^{N-n} P$ = the sum of all the weights from the end of the n-th bay to V_2; we have

$$S_s = V_1 - \Sigma_0^n p = -(V_2 - \Sigma_0^{N-n} p)\ldots\ldots(121)$$

and the stress will be $S_s \sec \theta$ ....(122)

It is easy to show, from Eq. (121), that the *primary maximum* is obtained for a uniform load, when the longer segment is loaded and the shorter unloaded.

88.—PROBLEMS *to be solved by Equation* (121).

1. Find the strain on the x-th pair of braces when the n-th receives its maximum for equal weights, the weight of the truss being included.

2. Suppose that N is odd and that equal weights are placed on the even joints, required the strain on the tie-braces between the n-th and $(n + 1)$ weights; the weight of the truss not considered.

3. Let $p_2 = 2p_1$; $p_3 = 3p_1$; $p_4 = 4p_1$, etc., $p_n = np_1$ re-

$$= \left\{ \left[(N-n')(N-n'+1)p + (N-2n'+1)Nw_1\right] \times \left[(-1)^{n'} \cos 2n'\pi\right.\right.$$
$$\left.\left. - \cos 2n'\pi\right] + \left[(N-n')(N-n'+1)p + (N-2n'+1)Nw_1\right] \times \left[(-1)^{n'}\right.\right.$$
$$\left.\left. \cos 2n\pi + \cos 2n'\pi\right] \right\} \frac{\sec\theta}{4N}$$

quired the shearing stress between the 1st and 2d load; the 3d and 4th; the $(n-1)^{th}$ and n^{th}. There being N bays in the lower chord, and the load on the lower chord.

4. Find the strain on the n-th tie-brace when all the joints in the upper and lower chords are equally loaded.

In this case the load may be equally distributed on both

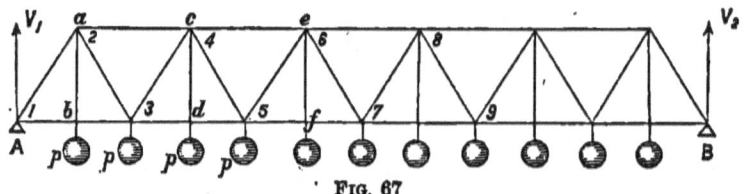

FIG. 67

chords, or it may all be on the lower chord, but in the latter case the weights at the middle of each bay must be supported directly by vertical ties which transmit the stress directly to the joints in the upper chord. The vertical ties are not essential parts of the truss. The strains are not transmitted through them as they pass from joint to joint. They simply transmit the load to a joint. In this case each of the essential tie-braces receives a different strain, and hence we find the strain upon the n-th tie-brace directly (not upon a pair). The braces are numbered in their order as shown in Fig. 67. *The strain on the n-th tie-brace is* *

$$(N - n + \tfrac{1}{2})\,p\,\sec\theta \dots\dots\dots\dots\dots (123)$$

---

* In this case we get a simple *modulus of signs*, for the signs will change with each successive piece. Calling tension + and compression —, and observing that the first brace is compressed and we have immediately —, the strain on the n-th tie-brace $= (-1)^n\,(N - n + \tfrac{1}{2})\,p\,\sec\theta$. If the truss be inverted, we have $- (-1)^n\,(N - n + \tfrac{1}{2})\,p\,\sec\theta$

Mr. J. Burkitt Webb, class of 1871, *University of Michigan*, was the first to suggest to me the manner of numbering shown in Fig. 67. By placing the figures in the angles as shown in the Fig., he makes them number the tie-braces; and also *the bays of* the chords at the same time.

If $N =$ the total number of tie-braces; $\tfrac{1}{2}\,N$ will be the number of bays in the lower chord; or if $l = \tfrac{1}{2}$ the length of a bay, the total span will be $L = Nl$.

Mr. Webb then gives the following equations:—

$$\frac{"N + 1 - 2n \mp \cos n\pi}{4}\,p\,\sec\theta\,\cos n\pi$$

5. Find the *maximum* strain on the n-th tie-brace when the live load is so placed as to produce a maximum, and the weight of the dead load is equal at all the joints of the upper and lower chords.

Ans. $\left[(2\ N - n)\ (2\ N - n + 1)\ p + (2\ N - 2\ n + 1)\ 2\ N w_1\right] \dfrac{\sec \theta}{4\ N}$ ....(124)

6. Find the strain on the x tie-brace in the preceding problem.

7. If the weights increase as the natural numbers from the first joint, that is, if the load on the first joint is p (this will not produce strains on the truss), on the second joint $2p$; on the third, $3p$; and so on, required the stress on the n-th tie-brace.

8. In the preceding problem, if successive weights are removed, beginning at 1, what will be the maximum strain on the n-th brace?

89.—STRAINS ON THE CHORDS.—*Strain on the supported chord.* In Fig. 64, let the bay $2-3$ be the n-th, conceive that this bay is severed, and instead of the strain, suppose that an equal stress, t_n, is substituted therefor. Taking the origin of moments at d, and the moment of V_1 will be $V_1 (n - \frac{1}{2}) l$; and of the $(n-1)$ p's, the moment will be $\frac{1}{2}(n-1)^2 pl$; hence if the weight of the truss is considered, we have

$$t_n = \left[(N-1)(p+w_1)(n-\tfrac{1}{2}) - (n-1)^2(p+w_1)\right]\dfrac{l}{2D}\ \ldots\ldots(125)$$

for the strain on the tie-braces, and

$$\dfrac{n(N-n) \pm \sin^2 n\dfrac{\pi}{2}}{4}\, p\tan\theta \sin\dfrac{2n-1}{2},$$

for the chords.

These give the strains on all the parts of a Warren girder for equal loads throughout. When the upper joints are loaded use the upper signs; when the lower joints are loaded use the lower signs; and when both upper and lower joints are loaded use both signs, that is, omit the ambiguous term and change the denominator to 2."

In a similar way we find that
The strain on the n-th bay of the upper chord is

$$c_n = (N - n) n (p + w_1) \frac{l}{2D} \dots\dots\dots\dots (126)$$

If $n = \tfrac{1}{2} N$, Eq. (126) gives $\tfrac{1}{8} \frac{WL}{D}$

90.—PROBLEM.—Find the stress on the n-th bay of the lower chord in Fig. 67. Also on the n-th bay of the upper chord.

91.—EXAMPLES. 1. Suppose that a triangular trussed bridge is 120 feet long clear span, each bay ten feet long, and the depth ten feet. Required the stresses due to a uniform dead load of sixty tons uniformly distributed over its whole length.

Here $L = 120$ feet; $N = 12$;
$l = 10$ feet; $p = 5$ tons;
$D = 10$ feet; sec. $\theta = 1.118$;
tang. $\theta = \tfrac{1}{4}$; No. of tie-braces = 24.

From Eq. (102) we have
The stress on the 1st and 24th................. − 33.54 tons.
The stress on the 2d and 23d.................. + 27.95 "
The stress on the 3d and 22d.................. − 27.95 "
The stress on the 4th and 21st................. + 22.36 "
The stress on the 5th and 20th................. − 22.36 "
The stress on the 6th and 19th................. + 16.77 "
The stress on the 7th and 18th................. − 16.77 "
The stress on the 8th and 17th................. + 11.18 "
The stress on the 9th and 16th................. − 11.18 "
The stress on the 10th and 15th................ + 5.59 "
The stress on the 11th and 14th................ − 5.59 "
The stress on the 12th and 13th................   0.00 "

Strains on the Chords.—Eq. (106) gives for the strain on the lower chord, 1st bay, 15 tons; 2d, 40 tons; 3d, 60 tons; 4th, 75 tons; 5th, 85 tons; 6th, 90 tons.

The stresses upon the bays of the upper chord are $27\tfrac{1}{2}$ tons; 50 tons; $67\tfrac{1}{2}$ tons; 80 tons; $87\tfrac{1}{2}$ tons; and if we conceive an exceedingly short bay at the middle, we have, by making $n = 6$ in the equation, 90 tons, the same as at the middle of the lower chord.

If the truss in the preceding example is made of wood whose weight is 50 lbs. per cubic foot, and the section of the parts proportional to the stresses, we may readily find that the weight of the truss will be about 7 tons; but if the braces are proportioned as columns, the weight will exceed this amount. The roadway and certain iron fastenings, etc., which are not included in the computation, would probably bring the weight up to 12 tons or more.

MINIMUM MATERIAL IN A TRIANGULAR TRUSS.   103

2. Let the length and depth of the truss be as in the preceding example. Assume that the weight of the truss is 15 tons (net), and the live load 60 tons, which is concentrated at the joints of the upper chord in equal amounts. Required the maximum strain upon the tie-braces as the load moves off without shock. Use Eq. (111) for the strains on the tie-braces.

Maximum stress on the

		First Diff.	Second Diff.
1st .. $-$ 41.93			
		6.76	
2d and 3d ... $\pm$ 35.17			ـ.47
		6.29	
4th and 5th... $\pm$ 28.88			0.47
		5.82	
6th and 7th... $\pm$ 23.06			0.47
		5.35	
8th and 9th... $\pm$ 17.71			0.47
		4.88	
10th and 11th... $\pm$ 12.83			0.47
		4.41	
12th and 13th... $\pm$ 8.42 (middle pair)			0.47
		3.9	
14th and 15th... $\pm$ 4.48 (secondary maximum)			0.47
		3.47	
16th and 17th... $\pm$ 1.01 (secondary maximum)			0.47
		3.00	
18th and 19th... $\pm$ 1.99			

Between the 17th and 18th the signs change, and the values after the 17th, which are found by the formula, are the amounts which the live load relieves the stresses which are due to the dead load.

3. Let $L = 60$ feet; $l = 5$ feet; $D = 10$ feet; $W_1 = 30$; $W_2 = 40$ tons on the upper chord. Required the maximum strains as the live load moves off without shock.

We have $N = 12$; $Nw_1 = 30$ tons; $w_1 = 2\frac{1}{2}$ tons; $p = 3\frac{1}{4}$ tons; $\sec\theta = 1.03$.

4. What must be the relation between the live and dead loads, both uniform and on the upper chord, so that the maximum strain on the end braces shall be x times the maximum strain on the central pair? N may be odd or even.

5. What must be the relation between the live and dead loads, so that the point of zero shearing shall be at xN; or $n_0 = xN$. x will be a fraction. Discuss the result, and show the limiting values of x.

92.—MINIMUM AMOUNT OF MATERIAL.—It is practically impossible to make a direct explicit solution of the problem for determining the minimum amount of material for a bridge of given span and load. The simple reason is—the functions are too complex. For instance, *consider a Warren girder in which the load is upon the upper joints, and N even.*

Let the ties be of iron, and the braces and tie-braces be of wood. The section of the former will be

$$k = \frac{F}{T}$$

in which $T =$ the modulus of tenacity, and
$F =$ the pulling strain.
The section of the latter will be

$$b^2 = l\sqrt{\frac{Q}{C}}$$

in which $l =$ the length of the piece;
$b =$ one side of the square section
$C =$ a constant depending upon the kind of material used; and
$Q =$ the pushing force.
The value of F, Eq. (111), or

$$F = \left[(N-n)^2 p + (N^2 - 2nN)w_1\right]\frac{\sec\theta}{2N}\ldots\ldots(127)$$

substituted in the value of k given above, and n made $= 1, 2, 3$, etc., to $\frac{1}{2} N$, will give a series of values which may be summed into a single expression.

The successive values of Q may be found from the same equation by making $n = 0, 1, 2$, etc., to $\frac{1}{2} N - 1$, which, substituted in the value of b^2 given above, gives another series which is not so easily summed. Near the middle of the truss we should determine whether a tie-brace when proportioned as a brace will not also serve as a tie, and if so it should not be considered in the preceding expression. It is thus seen, without any further considerations, that the problem is a very complex one.

Having found the sections of the pieces, we proceed to find the total volume. Thus, let

$L =$ the length of the span;
$l =$ the length of a bay;
$D =$ the depth of the truss; and
$\theta =$ the inclination of a tie-brace.

Then $l = \dfrac{L}{N}$; and $tang\,\theta = \dfrac{\frac{1}{2} l}{D}$

The length of a tie-brace $= D \sec \theta$; and the transverse sections as found above multiplied by $D \sec \theta$, and the results added will give the volume of the tie-braces. In a similar way the volume of the chords may be found, and the total volume is to be a minimum. But the problem is too complicated to make its solution of any practical value in this connection.

The problem becomes much simplified by supposing that the sections of the several pieces are directly proportional to the strains to which they are subjected. This case was solved by William Bouton, Class of 1865, *Univ. of Mich.* (See *Jour. Frank. Inst.*, Vol. LXXX., p. 80). As the hypothesis which he assumed is rarely realized in practice, I will not reproduce his solution, but will give in the following table some of the results of his analysis:—

TABLE

Giving the inclination of the tie-braces and the relation between the length and depth of a Warren girder for a minimum amount of material when the sections are proportional to the stresses.

VALUES OF		θ	SPAN ÷ DEPTH.
N	$\frac{w_1}{p}$		
6	½	19° 11′	4.17
9	⅓	16° 14′	5.03
12	0	14° 37′	6.25
40	½	8° 8′	11.43
50	0	7° 28′	13.11

The problem which has usually been given under this head is—*To find the inclination of the braces and ties so as to give a minimum amount of material, the sections being proportional to the stresses to which they are subjected.*

In this case, using the same notation as before, we have
  $K =$ the section of any brace or tie;
  $D \sec \theta =$ the length of a brace; and
  $L \div D \tang \theta =$ the number of tie-braces.

$$\therefore \frac{KDL}{D}\frac{\sec\theta}{\tan g\,\theta} = \text{the volume of all the braces, which is to}$$

be a minimum. From Eq. (118) we see that K varies as $\sec\theta$, and also with n, but in determining the inclination we are not concerned with the latter; therefore the essentially variable part of the above expression becomes

$$\frac{\sec^2\theta}{\tan g\,\theta} = \frac{1}{\cos\theta \sin\theta}$$

which is a minimum for $\theta = 45°$.

But when the braces are proportioned as columns, this solution becomes of no practical value.

93.—PROPER LENGTH OF THE BAYS.—The length of the bays is, within certain limits, an arbitrary quantity. If too long they must be proportioned to resist a transverse strain as well as tension and compression. After the engineer has fixed upon the length of the bay and the depth of the truss, he may find that a single system of trussing will give too small an inclination to the tie-braces for economy; in which case a double, triple, or multiple system may be used, as shown in the following cases.

94.—CASE V.—DOUBLE TRIANGULAR SYSTEM WITH THE LOAD ON THE JOINTS OF THE SUPPORTED CHORD.

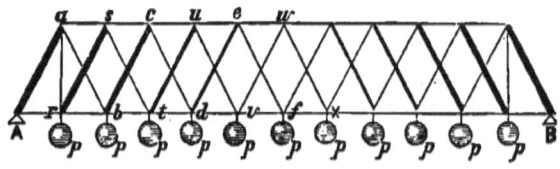

FIG. 68.

This is sometimes called a HALF LATTICE.

95.—STRESS ON THE TIE-BRACES.—The proper mode of analyzing this truss is to consider it as composed of two simple triangular trusses. The truss $A\ a\ b\ c\ d\ e\ f$, etc., is the same as Fig. 64, with the load on the supported chord, and the system $A\ a\ r\ s\ t\ u\ v\ w\ x$, etc., is essentially the same as Fig. 62, since in both cases the first load is one-half a bay from the supported end. Hence, the strains in the former system are given by Eq. (120) and in the latter by Eq. (111). The end brace $A\,a$ is common to both systems, and the stress on it is $\frac{1}{2}\,\Sigma p\,\sec\theta$.

96.—STRESS ON THE CHORDS.—The stress upon the several bays of the chords may be found by treating the two trusses separately, as in the preceding article. Thus, to find the stress on td, *first* find it on bd of the system $A\,abcde$, etc., which will be given by Eq. (125); *secondly*, find it on tv of the system $A\,arstuv$, etc., which will be given by Eq. (106), and *thirdly*, add the results for the stress on td; the part which is common to both bd and tv.

97.—TOWNE'S LATTICE is a multiple triangular system, as shown in Fig. 69. It was designed many years since, long before iron was used in this country, for bridges. It is composed of planks of uniform thickness and width, placed at equal distances from each other, and having equal inclinations in opposite directions. The planks are secured to each other at their crossings by wooden pins—sometimes called tree-nails.

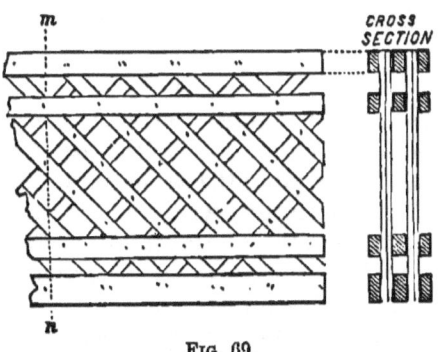

Fig. 69.

The chords are also formed of planks of the same size as the lattice-work, and arranged so as to break joints. If the system is doubled, they are separated by one of the stringers of the chord, as shown in the **"CROSS SECTION"** in the figure. The construction is evidently very simple, and may be made by an ordinary carpenter. They may be made by the mile, and cut off by the yard to suit the occasion. Notwithstanding their simplicity they have proved themselves to be very efficient in supporting loads. A railroad bridge on this plan has been in use for many years over the Hudson River, at Troy, N. Y., and is now in good condition. Trusses on this plan have been made of flat strips of iron, which were riveted together at their crossings. Samples of such structures may be seen at Schenectady and Rome, N. Y., on the New York Central Railroad.

98.—ANALYSIS OF TOWNE'S LATTICE.—Since the chords are of uniform size throughout, we have only to determine the proper size for them at the middle of the span. Suppose that the load is uniformly distributed over the whole length.

Let W = the total uniform load including the weight of the bridge;
L = the span;
D = the depth of the truss; and
H = the pull at the middle of the lower chord, or compression at the middle of the upper chord.

Then

$$HD = \tfrac{1}{2} W \tfrac{1}{2} L - \tfrac{1}{2} W \times \tfrac{1}{4} L = \tfrac{1}{8} WL.$$

$$\therefore H = \frac{WL}{8D}$$

If there are two chords this stress will be divided by two, and the quotient used for determining the size of the chords.

To determine the strength of the lattice-work, let a vertical section, as $m\,n$, Fig. 69, be made near the end of the truss, and let

m = the total number of tie-braces cut by the vertical plane in all the trusses which carry the given load;
k = the section of each tie-brace;
$C = T$ = the modulus of strength; and
θ = the inclination of the tie-braces.

Then

$m\,Tk\cos\theta$ will equal the vertical shearing stress.

$$\therefore m\,T k \cos\theta = \tfrac{1}{2} W$$

$$\therefore m = \frac{W}{2\,kT\cos\theta}$$

LATTICE BRIDGES. 109

This is an approximate solution, but it is sufficiently exact for a wooden truss of this kind, in which the trussing and chords are uniform throughout; but in the case of an iron lattice, where it would result in a saving if the parts were properly proportioned, a more exact analysis is desirable, which may be made as in the following article.

99.—ANALYSIS OF THE MULTIPLE TRIANGULAR SYSTEM.—Let there be four systems of triangular trusses, as in Fig. 70. We here consider the effect on each system, as if the

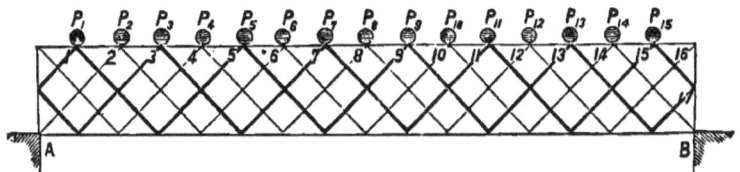

FIG. 70.

others did not exist, and then conceive that they are combined by being placed side by side so that their spans will coincide. Thus, beginning with p_1, we see that p_1, p_5, p_9, and p_{13} are carried by one system. Similarly, p_2, p_6, p_{10}, and p_{14} are carried by another system, and so on. The stress on each diagonal due to each weight may easily be found by the methods already explained. The diagonals which incline to the right in the figure are numbered from 1 to 17 inclusive.

EXAMPLE.—Let the span be 120 feet, depth of truss 15 feet, and 4 systems of right-angled triangles, as in Fig. 70. Let the permanent load due to the weight of the bridge be $2\frac{1}{4}$ tons at each apex, and the moving load 5 tons at each apex.

$$\text{Then } p_1 = p_2 = p_3 = \text{etc.}, = 5 \text{ tons};$$
$$w_1 = 2\frac{1}{4} \text{ tons};$$
$$N = 16 = \text{the number of bays in the span; and}$$
$$\theta = 45°.$$
$$\therefore \sec \theta = 1.414$$

Consider p_1. The horizontal distance between it and A is *one* bay; and between it and B is 15 bays; hence, A carries $\frac{15}{16} p_1$ and $B \frac{1}{16} p_1$; or A carries $\frac{75}{16}$ of a ton and $B \frac{5}{16}$ of a ton. The strain on the brace numbered 1 will be $\frac{75}{16} \times 1.414 = 6.625$ tons. Retaining only one decimal figure, and the result, 6.6, is entered in the following table.

110   TREATISE ON BRIDGES.

No. of the Diagonal	p_1	p_2	p_3	p_4	p_5	p_6	p_7	p_8	p_9	p_{10}	p_{11}	p_{12}	p_{13}	p_{14}	p_{15}	Max. $-p$	Max. $+p$	$w \pm$	Total $-$	Max. $+$
1	−6.6	−6.9	−5.7	−5.3	−4.9	−4.4	−4.0	−3.5	−3.1	−2.7	−2.2	−1.8	−1.3	−0.9		−15.9		−7.9	−22.8	
2	+0.4														−0.4	−14.2		−7.1	−21.3	
3		+0.9														−12.3		−6.1	−18.4	
4	+0.4		+1.3	−5.3	−4.9	−4.4	−4.0	−3.5	−3.1	−2.7	−2.2	−1.8	−1.3			−10.6		−5.3	−15.9	
5		+0.9												−0.9	−0.4	−9.3	+0.4	−4.4	−13.7	
6				+1.8												−8.0	+0.9	−3.5	−11.5	
7	+0.4		+1.3		+2.2	−4.4	−4.0	−3.5	−3.1	−2.7	−2.2	−1.8				−6.6	+1.3	−2.6	−9.2	
8		+0.9				+2.7								−0.9	−0.4	−5.3	+1.8	−1.7	−6.0	
9				+1.8			+3.1									−4.4	+2.6	−0.9	−5.2	
10	+0.4		+1.3		+2.2			+3.5	+4.0	−2.7	−2.2					−3.6	+3.6	0.0	−3.6	
11		+0.9				+2.7				+4.4				−0.9	−0.4	−2.6	+4.4	+0.8	−1.8	
12				+1.8			+3.1		+4.0		+4.9					−1.8	+5.3	+1.7	−0.1	
13			+1.3		+2.2			+3.5				+5.3				−1.3	+6.6	+2.6		+0.1
14		+0.9				+2.7										−0.9	+8.0	+3.5		+1.8
15				+1.8			+3.1						+5.7			−0.4	+9.3	+4.4		+3.6
16																	+10.6	+5.3		+5.2
17																	+12.3	+6.1		+7.0

LATTICE BRIDGES.   111

The stress on the braces — 5, 9, 13, and 17 due to p_1 is the same on each, and is $\rlap{/}{8} \times 1.414 = (\rlap{/}{4}$ of that on brace 1) $= 0.44\frac{1}{4} +$. Retaining only one decimal as before, and the results are entered in the preceding table.

The stress on brace 2, due to p_2 is $\frac{1}{8} p_2 \times 1.414 = 0.186$, which call 0.2 as in the table. Proceed in this way with each of the weights, and enter the results in the table.

In this example, + is for tension, and — for compression. The maximum tension is found by adding all the + values in each horizontal line, and the maximum compression by adding all the corresponding — values. These results are given in the columns marked "Max. — p" and "Max. + p." The values of the strains due to the weight of the bridge, w_1, are evidently one-half the strains which would result from loading every joint in the upper chord with p; and hence may be found by taking one-half the algebraic sum of the + p's and — p's. The results are entered in the column marked + w_1.

The strains in the columns + p and — p are those due to a live load; those for — p being the strains due to a load moving off to the right, and those of + p due to the same load moving off to the left. Those in the column + w_1 are the strains due to a permanent uniform load.

The maximum strains due to both the *live* and dead loads are evidently the algebraic sum of the strains due to each acting separately. These results are entered in the two last columns, the *minus* values being the strains due to the live load moving off to the right when the weight of the bridge is considered, and the other the effect of moving to the left.

100.—AMBIGUITY IN REGARD TO STRAINS IN CERTAIN CASES.—It has been shown, in article 71, that when the joints are equally loaded and symmetrically placed in reference to the centre, that all the tie-braces between the two most central pair of loads receive no strains. If this were generally true the strain due to the weight p_1 on the brace at the right of p_1, would be balanced by the strain on 9 due to the equal weight p_2; in which case neither would receive any stress due to an uniform load throughout. But an examination of the figure shows that the stress on 9 due to p_2 does not pass through the system of which p_1 is a member. If the truss were only half as deep, the two tie-braces mentioned would belong to the same system. An ambiguity here arises as to which way the strains will definitely pass; whether they will pass wholly by their own system, or partly by one and partly by the other. If proportioned according to an exact analysis, that hypothesis should be assumed which will give the greatest stress on any particular piece. This case is analogous to that of a rigid beam or frame which is supported by four or more props, in which case it is impossible to tell exactly how much is supported by each prop If supported by three props, it is easy to tell the amount sustained by each.

101.—WARREN'S GIRDER MODIFIED.—Returning now to the more simple form of the triangular truss with equally inclined tie-braces, if we conceive that the load remains on the supported chord, as in Fig. 65, and that the upper joints are carried forward any amount, as for instance, until one-half the tie-braces are vertical, as in Fig. 71, it is evident from the way

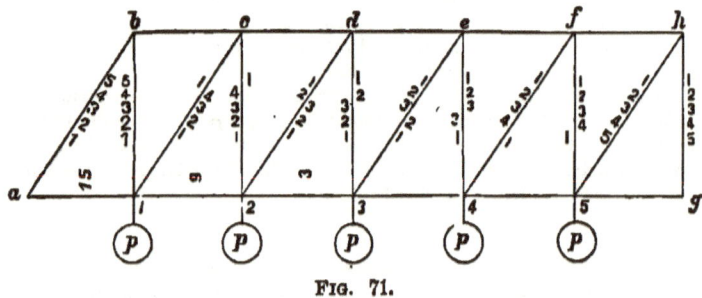

FIG. 71.

that the change has been made, as well as from an examination of the *distributed strains*, that the only difference in the two cases is in the secant of the inclination of the pieces, and hence Eq. (120) is applicable to this form.

If the joints are loaded in any manner, the more general case is shown by Fig. 72, in which t denotes tension and c compression.

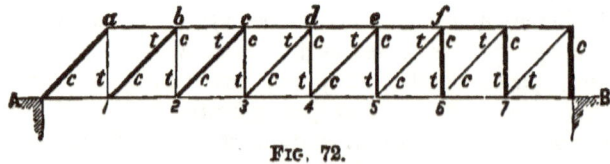

FIG. 72.

If now the right-hand half of the upper joints of Fig. 71 be moved to the left, so as to bring e indefinitely near d, we shall have the case shown in Fig. 73, which is shown in a still more general way in Fig. 74, in which the numbers are omitted and the character of the strains is indicated by the letters c and t.

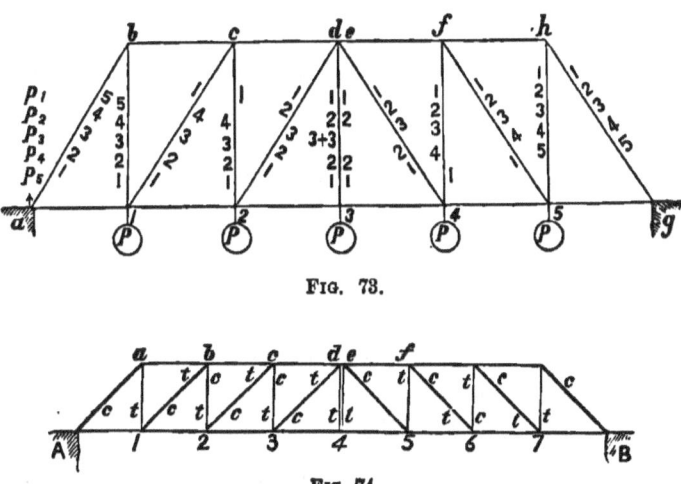

FIG. 73.

FIG. 74.

This is also a triangular truss, and to which Eq. (120) is directly applicable. If every joint be loaded, as in Fig. 75, then the inclined pieces will be subjected to compression only; but if there be an eccentric load, or in other words more load on one side of the centre than on the other, and the inclined pieces are arranged and secured so as to resist only compression (or tension), the truss may become distorted, as shown in Fig. 76.

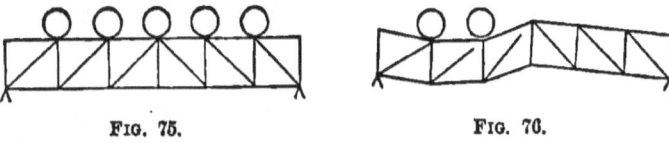

FIG. 75.   FIG. 76.

This kind of distortion is prevented by the arrangement of the *panel* system, which will now be explained.

TYPE FORM OF HOWE'S TRUSS.

102.—If now we introduce braces in each of the diagonals which incline in the opposite direction to those in Fig. 73, to receive by compression the strains of tension which, in Fig. 73, fall upon the inclined pieces, we shall have the form shown in Fig. 77.

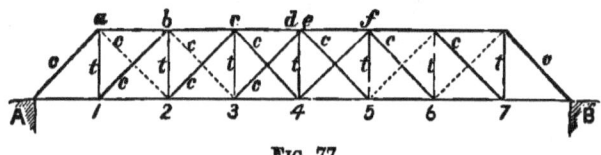

FIG. 77.

By this arrangement *tie-braces* are entirely eliminated, for it will be observed that all the inclined pieces are *braces*, being subjected to compressive strains only; and the vertical ones to tension only.

The TYPE FORM of this style of truss is called the **HOWE TRUSS.**

TYPE FORM OF PRATT'S TRUSS.

103.—In Fig. 72, if the joints of the left half of the upper chord be moved outward so as to make the end brace vertical, we shall have the form shown in Fig. 78, which is also a

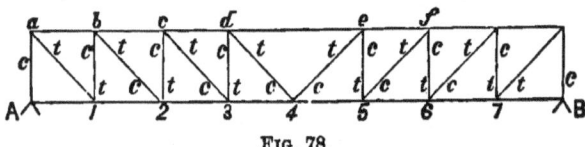

FIG. 78.

triangular truss, and to which Eq. (120) is directly applicable.

If now we introduce diagonal rods to resist by tension the stress of compression, which, in Fig. 78, act upon the tie-braces, we shall have the form shown in Fig. 79, in which all the inclined pieces are *ties*, and hence the vertical ones are struts;

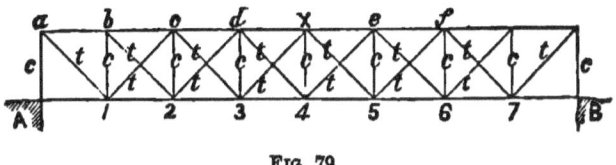

FIG. 79.

and hence in *this case we have no tie-braces*. The **TYPE FORM** of this truss is called a **PRATT TRUSS**.

PANEL SYSTEM.

104.—The Howe and Pratt trusses come under the more general head of the PANEL SYSTEM. The *panel system* consists of a system of trussing in which a part of the pieces are vertical or nearly so; and of inclined pieces so arranged that neither are subjected to opposite strains. As the latter principle applies to cases in which the chords are not parallel, and in which the vertical members (so called) are somewhat inclined, the still more general term of QUADRANGULAR SYSTEM is sometimes used.

A *panel* is one of the rectangular (or quadrangular) spaces 1 *b c* 2, Fig. 79, included between two verticals and the chords.

A *diagonal* is an inclined piece which crosses a panel obliquely. The *diagonal* in the "Howe Type" is a *brace;* and in a "Pratt Type" it is a *tie.*

The *verticals* in the "Howe Type" are *ties*, and in the "Pratt Type" they are *struts* or *posts.*

105.—MAXIMUM STRESS ON THE DIAGONALS.—PROPOSITION.—WHEN THE VERTICAL MEMBERS ARE EXACTLY VERTICAL, THE STRESS ON THE DIAGONALS IS THE SAME WHETHER THE LOAD BE UPON THE UPPER OR LOWER CHORD, OR UPON BOTH.

In Fig. 80, for instance, if a weight rests at 3, and *b* 3 is a tie, that portion of the weight which is sustained by the support at *A* will be transmitted through the tie *b* 3. But if *b* 3 is a brace it cannot transmit this stress, but instead thereof it will be first transmitted to c, and thence to A through the brace $c\,2$. The same reasoning applies if the load is placed at c. Hence, A will carry the same amount whether the load be at 3 or c.

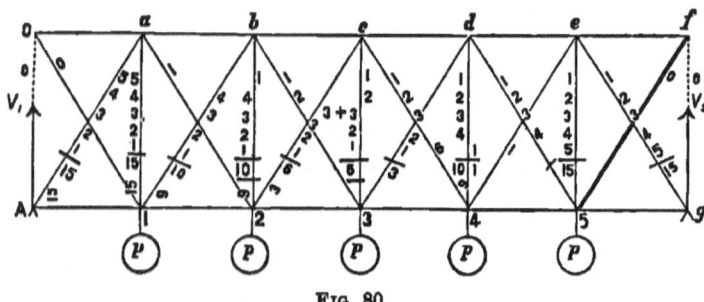

FIG. 80.

Or, more generally, we observe that the vertical forces between $b\,2$ and $c\,3$ are transmitted from one side of the panel to the other through the diagonals, and hence it makes no difference where the load is placed, provided the points of application are in the same vertical.

An examination of Fig. 80 shows that to produce a maximum stress on any diagonal, the truss must be loaded from that diagonal to the end towards which the diagonal acts. For example, if $b\,3$, Fig. 80, is a brace, the maximum stress on it will be induced by a load from 2 to A; but if $c\,2$ is a brace, the maximum stress will be induced by a load from 3 to g. If $c\,2$ were a *tie*, the maximum stress on it would be induced by a load extending from 2 to A.

Let $N =$ the number of bays in the chord;
$\quad n =$ the number of the brace or tie considered, counting from either end. This will also equal the number of bays in the supported chord between the end and the brace (or tie) which is being considered, including the bay which is in the same panel as the brace (or tie); it also represents the part which has no live load;
$\quad p =$ one of the equal weights of the live load;
$\quad w_1 =$ one of the equal weights of the dead load;
$\quad \theta =$ the inclination of a brace (or tie); and

PANEL SYSTEM. 117

$V_1 =$ the reaction of the support at the end from which n is counted.

Then, if it is so loaded with equal weights at the joints, as to produce a maximum strain upon the n-th brace (or tie) we have

$$V_1 = \frac{(N-n)(N-n+1)}{2N} p + \tfrac{1}{2}(N-1) w_1.$$

Subtracting $(n-1) w_1$ from this, which is all the load between the end and the n-th diagonal, and we have

$$S_n = \left[(N-n)(N-n+1) p + (N-2n+1) N w_1\right] \frac{1}{2N},$$

which, multiplied by the $\sec \theta$ gives for

The maximum stress on the n-th brace (or tie) of the panel system for equal weights of live and of dead loads, the load being either on the upper or lower chord, or on both:—

$$F_n = \left[\begin{array}{l}(N-n)(N-n+1) p \\ + (N-2n+1) N w_1\end{array}\right] \frac{\sec \theta}{2N} \ldots\ldots\ldots(128)$$

which is the same as Eq. (120).

106.—DISCUSSION OF EQUATION (128).

1. Let $N = 2$, and $n = 1$, and Eq. (128) becomes

$$\tfrac{1}{2}(p + w_1) \sec \theta.$$

This is the case of the king-post truss, and the preceding Equation is substantially the same as the second of Eqs. (68).

2. Let $N = 3$, the case of the queen-post or trapezoidal truss, and we have

$$F_n = \left[(3-n)(4-n) p + (4-2n) 3 w_1\right] \frac{\sec \theta}{6}$$

in which, if $n = 1$, we have

$$F_1 = (p + w_1) \sec \theta$$

which is substantially the same as Eq. (96).

If $n = 2$, we have

$$F_2 = \tfrac{1}{3} p \sec \theta.$$

in which w_1 disappears, as it should. This is the case in which a weight p is placed at one of the joints, and the other one is unloaded as in Fig. 51, and $\frac{1}{2} p \sec \theta$ is the stress which would fall upon a diagonal CF, Fig. 49.

3. If $n = 1$, we have from (128)

$$F_1 = \tfrac{1}{2}\Big[(N-1)(p + w_1)\Big] \sec \theta. \quad \ldots \ldots \ldots \ldots (129)$$

And observing, that for a load *uniformly distributed* over the whole length, each support would sustain $\tfrac{1}{2} N (p + w_1)$, and it appears that the end braces carry less than one-half the total load. This is because one-half the load on each of the end bays, when the load is uniformly distributed, is carried directly by the abutments, and hence the strains due to this part of the load are not transmitted through the trussing. This shows a distinction between a load uniformly distributed and one composed of equal weights placed at the joints, and will be noticed hereafter.

4. For equal weights throughout, the term

$$\tfrac{1}{2}(N - 2n + 1) w_1 \sec \theta \ldots \ldots \ldots \ldots (129\ a)$$

gives the strains on all the diagonals; and if $n = \tfrac{1}{2}(N + 1)$, (129-a) reduces to zero, which shows that for an odd number of bays and uniform load there will be no strain on the braces in the central panel. See Fig. 49.

5. If the weight of the truss be neglected, $w_1 = 0$, and

$$F_n = (N - n)(N - n + 1) p \frac{\sec \theta}{2 N},$$

gives the stress on the n-th diagonal. This becomes zero for $n = N$, or $n = N + 1$; the later value of which being beyond the truss, does not belong to the practical part of the problem.

6. For n less than $\tfrac{1}{2}(N + 1)$ both terms of Eq. (128) are positive, and hence when the live load extends over more than half the length of the truss, the dead and live loads conspire to produce strains.

7. For n greater than $\tfrac{1}{2}(N + 1)$, the first term (containing p) remains positive, but the second term (containing w_1) becomes negative; and hence, when the shorter segment only is loaded the live and dead loads act against each other in producing strains.

STRESS ON THE DIAGONALS. 119

8. If we place Eq. (128) equal to zero, and solve for n, we find

$$n = \frac{w_1}{p} N + N + \tfrac{1}{2} \pm \sqrt{\left(\frac{w_1^2}{p^2} + \frac{w_1}{p}\right) N^2 + \tfrac{1}{4}}. \quad (130)$$

which gives the brace on which there is no strain. To make the nature of this equation clear, suppose that a live and dead load, consisting of equal weights, is placed at every joint of the upper or lower chord, in which case the stress on the first brace is a maximum for that load. Then remove the weight at one end which represents the live load, and we shall have the maximum stress on the second brace. Remove the second weight of the live load, and we have the maximum stress on the third brace, and so on; the stress on each successive brace being a maximum although less than that on the preceding until, after passing the middle, we find a brace which has no stress, or which has a negative stress. The exact point at which the stress changes signs is given by Eq. (130). In other words, if we conceive a live load moving off from a bridge, the braces just under the rear end of the load will receive their maximum stress for that load, and the braces under action will incline towards the rear end of the load; but there will be a point at which there will be no stress; and beyond that point no braces which incline towards the load are necessary. This is illustrated by Fig. 81. The peculiarity shown in the chords will be noticed hereafter. The strains due to the live load after it passes N_o (see next page) will simply relieve the remaining braces of a portion of the strain which is due to the permanent load.

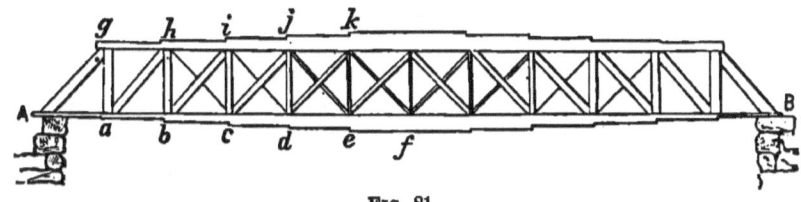

FIG. 81.

There are two values of n in Eq. (130), one of which exceeds $\tfrac{1}{2}(N+1)$ and is less than N, and the other value exceeds N, and hence does not belong to the practical part of the problem. That value has only a theoretical signification. Call the former value n_o. We also find, by assuming values for N that the value of n_o is generally fractional; but as n is the number of a brace, and as there are no fractional braces, we retain only the integral part of n_o. If n_o is an integer, it shows that there is no stress on the n-th brace (it is neutral), and hence practically we only need $n_o - 1$.

Let $n_o =$ that value of n in Eq. (130), which is $< N$ and $> \frac{1}{2}(N + 1)$; and

$N_o =$ the integral part of n_o; or if n_o is an integer $= n_o - 1$; then

$N_o =$ the number of braces (or ties) which should incline one way counting from the end; and

$N - N_o =$ the number of panels in which only one brace (tie) is needed, as in Fig. 81.

107.—A COUNTER-BRACE is one which inclines in the opposite direction from a main brace. The main braces incline *from* the end. The counter-braces incline from the middle. An examination of Fig. 81 shows that the counter-braces beyond the middle (counting from either end) incline the same way as the main braces between the end and middle; and the analysis in the preceding article shows that the formula which is applicable to main braces is also applicable to counter-braces. According to this mode of reasoning, a counter-brace possesses no peculiarity which is not common to a main-brace, and hence the term *counter-brace*, in an analytical sense, is unnecessary but in practice its use is common. Still using the term, we have, analytically,

$N_o - \frac{1}{2}(N + 1) =$ the number of panels, each side of the centre, which require counter-diagonals, and hence on both sides of the centre including the central one we have

$2 N_o - N =$ the number of panels in each truss in which diagonals should incline both ways; and

$2 (N - N_o) =$ the number of panels which require main diagonals only; and

$N - N_o =$ the number of panels near each end which do not require counter-diagonals.

108.—VALUES OF N_o.

Table.— Values of N_o, or the number of braces (or ties) which should incline from either end in the Panel System.

N	$w_1 = 0$	$p = 10\, w_1$	$p = 5\, w_1$	$p = 2\, w_1$	$p = w_1$	$p = 1.5\, w_1$	$p = 0$
3	2	2	2	2	2	2	1
4	3	3	3	2	2	2	2
5	4	4	3	3	3	3	2
6	5	5	4	4	3	3	3
7	6	5	4	4	4	4	3
8	7	6	5	5	5	4	4
9	8	7	6	6	5	5	4
10	9	8		6	6		5
12	11	9	8	8	7	6	6
15	14	11	11	9	9	8	7
20	19	15	14	13	12	10	10
30	29	23	21	19	18	16	15
40	39	31	28	25	23	21	20
50	49	38	35	32	29	26	25

Number of panels near each end which need no counter-diagonals.

N	$w_1 = 0$	$p = 10\, w_1$	$p = 5\, w_1$	$p = 2\, w_1$	$p = w_1$	$p = 1.5\, w_1$	$p = 0$
3	1	1	1	1	1	1	2
4	1	1	1	2	2	2	2
5	1	1	2	2	2	2	3
6	1	1	2	2	3	2	3
7	1	2	3	3	3	3	4
8	1	2	3	3	3	4	4
9	1	2	3	3	4	4	5
10	1	2	4	4	4	4	5
12	1	3	4	4	5	6	6
15	1	4	4	6	6	8	8
20	1	5	6	7	8	10	10
30	1	7	9	11	12	14	15
40	1	9	12	15	17	19	20
50	1	12	15	18	21	24	25

109.—EXAMPLES.—1. In a truss of the Panel System, let the clear span be 120 feet, and the depth from centre to centre of chords be 10 feet; and the length of each bay be 10 feet; total weight of the truss 60 tons; live load 60 tons when it extends over the whole length of the truss. Required the maximum stress on the diagonals as the live load moves off without shock. (We assume in this that the live load is equivalent to equal weights placed at the joints.)

We have $p = w_1 = (\frac{66}{12} = 5)$ tons $= 10,000$ lbs.

$N = 12$; $\sec \theta = \sec 45° = \sqrt{2} = 1.4142$.

These values in Eq. (128) and n made equal to 1, 2, 3, etc., give:—

	First Diff.	Second Diff.
Stress on 1st brace = 155,562 lbs.		
	27,106	
Stress on 2d brace = 128,456 lbs.		1,178
	25,928	
Stress on 3d brace = 102,528 lbs.		1,178
	24,750	
Stress on 4th brace = 77,778 lbs.		1,178
	23,572	
Stress on 5th brace = 54,206 lbs.		1,178
	22,394	
Stress on 6th brace = 31,812 lbs.		1,178
	21,216	
Stress on 7th brace = 10,596 lbs.		
Stress on 8th brace = (negative)		

Hence n_o is between 7 and 8, and $N_o = 7$ in this example; and hence only one counter-brace is required according to this series; and only two in the whole truss.

2. The following are the actual dimension of a bridge truss. Length = 100 feet, $N = 8$ $\therefore l = 12\frac{1}{2}$ feet. $D = 19$ feet. $\therefore \sec \theta = 1.2$. Weight of bridge $W_1 = 60$ tons; uniform live load $W_2 = 100$ tons (net). $\therefore p = 12\frac{1}{2}$ tons, and w_1 $7\frac{1}{2}$ tons.

There were two trusses, but by computing as if there were but one, we find for the maximum stress as follows:—

On the 1st diagonal, 84 tons; 2d, $61\frac{3}{4}$ tons; 3d, $41\frac{4}{5}$ tons; 4th, $23\frac{1}{2}$ tons; 5th, $6\frac{1}{4}$ tons; and on the 6th it is negative; hence three panels need no counter-diagonals.

3. The following are the dimensions of a whipple bridge (see Fig. 89) on the Albany and Rutland Railroad near Troy, N. Y.

Total length 147 feet; clear span, $L = 145$ feet; $N = 14$ $\therefore l = 10\frac{1}{4}$ feet; $D = 21\frac{1}{2}$ feet; $\sec \theta = 1.1$. Total weight of the bridge $61\frac{1}{4}$ tons, $\therefore w_1 = 4.4$ tons. Assume that the live load is one ton per running foot; and hence $p = 10\frac{1}{4}$ tons. Required the maximum strains on the ties. (The truss is also 17 feet wide from out to out; suspension ties 31 feet. There were two trusses.)

4. If in a panel truss $N = 12$, and it is found that for a live and dead load, the maximum stress on the 3d diagonal is 21.9 tons; and on the 4th it is 16.8 tons; $\sec \theta = 1.2$; required the values of p and w_1, and the total weight of the truss. $Ans.$ $W_1 = 24$ tons.

5. If $N = 10$ and $p = 12$ tons, and it is found that the strain on the 7th brace is zero when the live load is on the three end joints; required w_1 or the weight of the bridge.

6. I have used for illustration a model bridge of 12 panels which weighed 19 lbs., and I found that the maximum stress on the fourth brace for

STRESS ON THE DIAGONALS.   123

equal weights on 8 end joints was 21 lbs.; $sec\ \theta = 1.4$. Required the value of one of the equal weights, p.

7. Suppose that a panel truss is so loaded with equal weights as to produce maximum strains on the following diagonals:—on the second diagonal, 5,448 lbs.; on the fourth, 4,208 lbs.; and on the tenth, 680 lbs. If $sec\ \theta = 1.6$, required the number of panels in the truss; the weight of the bridge, and the weight of the live load.

8. Two trusses each weigh 10 tons, $W_2 = 20$ tons on each; one has 10 panels, the other 20 panels; the total length and $sec\ \theta$ the same in both. Required the stress on the fifth brace of the former and tenth of the latter; and show why they are not the same.

110.—GENERAL VALUE OF THE SECOND DIFFERENCES.
—In Eq. (128) make $n = 1$, and we have

$$\left[\frac{(N^2 - N)p}{2N} + \tfrac{1}{2}(N - 1)w_1 \right] sec\ \theta$$

For $n = 2$, we have

$$\left[\frac{N^2 - 3N + 2}{2N}p + \tfrac{1}{2}(N - 3)w_1 \right] sec\ \theta$$

For $n = 3$, we have

$$\left[\frac{N^2 - 5N + 6}{2N}p + \tfrac{1}{2}(N - 5)w_1 \right] sec\ \theta$$

The difference between the second and first of these values is

$$\left[\frac{2N - 2}{2N}p + w_1 \right] sec\ \theta$$

Between the second and third, it is

$$\left[\frac{2N - 4}{2N}p + w_1 \right] sec\ \theta$$

These are the first two terms of the first differences. The difference between these is

$$\frac{p}{N} sec\ \theta$$

which is the constant value of the second differences.

In the first example above, this becomes

$$\frac{5 \times 2,000 \times 1.4142}{12} \text{ lbs.,} = 1,178\tfrac{1}{2} \text{ lbs.}$$

124   TREATISE ON BRIDGES.

The first term of the first difference, as given above, is

$$\left[\left(1 - \frac{1}{N}\right)p + w_1\right] \sec \theta = \left[\frac{11}{12} \times 10{,}000 + 10{,}000\right] \times$$

$1.4142 = 27{,}106$ lbs. The successive terms in the column of first differences may be found by the successive subtractions of the second differences $= 1{,}178$ lbs. (omitting the $\frac{1}{2}$ lb.).

The stress on the first brace is $\frac{1}{2}(N-1)(p+w_1)\sec\theta = \frac{1}{2} \times 11 \times 10 \times 2{,}000 \times 1.4142 = 155{,}562$, and by subtracting, successively, the successive first differences, the actual strains may be found.

111.—STRESS ON ANY DIAGONAL.—REQUIRED THE STRESS ON THE x-TH DIAGONAL WHEN THE TRUSS IS SO LOADED WITH LIVE AND DEAD LOADS AS TO PRODUCE A MAXIMUM ON THE n-TH DIAGONAL.

Let $N =$ the total number of bays in the supported chord;
  $p =$ one of the equal weights of the live load;
  $w_1 =$ one of the equal weights of the dead load;
  $\theta =$ the inclination of a diagonal;
  $n =$ the number of a diagonal which receives the maximum stress; and
  $x =$ the number of the diagonal which is considered.

There are two cases :—

  1st, for $x < n$, and
  2d, for $x > n$.

1st, For $x < n$ we find the stress, by deducting from the value of V_1, which is given just before Eq. (128), all the load between the end and the x-th brace, which is $(x-1)w_1$ and multiplying the remainder by $\sec \theta$.

$$\therefore F_x = \frac{\left[(N-n)(N-n+1)p + (N-2x+1)Nw_1\right]\sec\theta}{2N} \quad \ldots (131)$$

2d, For $x > n$ we use the same principle, but the load to be deducted will be $(x-1)w_1 + (x-n)p$; hence we have

$$F_x = \left[\big((N-n)(N-n+1) - 2N(x-n)\big)p + (N-\right.$$

$$2x + 1)\, N\, w_1 \Big]\, \frac{\sec \theta}{2\, N} \quad \dots\dots\dots\dots\dots\dots\dots\dots \quad \dots (132)$$

These equations may be discussed in the same way as Eq. (128), but they are not of much practical value, since the parts should always be proportioned to resist the *maximum* stress.

EXAMPLE.—If a bridge has 10 bays, and the permanent load on each joint is $w_1 = \tfrac{1}{4}$ ton, and the transient load $p = 1$ ton, required the stress on all the braces when the third receives the maximum stress.

112.—UNIFORMLY DISTRIBUTED LOAD.—A load uniformly distributed over a portion of the span, as in Fig. 82, is not the same as if equal weights were placed upon the truss

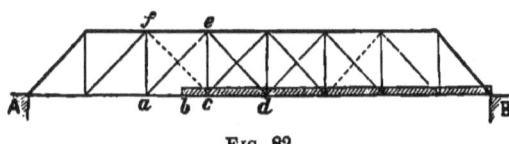

FIG. 82.

from c to B. We are considering the joints as perfectly flexible; hence the joint c carries one-half of the load on $c\,d$, but only a fractional part of that on $a\,c$; and the other part on $a\,c$ is carried by a. If the load extends from a to B, then the load on c is one-half that on $a\,c$ and $c\,d$, and that on a is one-half that on $a\,c$. Hence, it appears that it is impossible to produce equal weights at the joints for a uniform load, except when the load extends over the whole length of the truss. Let us therefore ascertain how far the load must extend to produce a maximum stress on the n-th diagonal, which, in Fig. 82, is the diagonal $a\,e$.

Let N, n, and p, be the same as before;
$w =$ the load per foot of length;
$x = b\,c \therefore w\,x =$ the load on $b\,c$; and
$l = a\,c$.

$\therefore$ the joint a sustains directly $\tfrac{1}{2}\dfrac{w\,x^2}{l}$, and c sustains

directly $\dfrac{w\,x(l - \tfrac{1}{2}x)}{l}$.

It is evident that the maximum stress on ae is the difference of the stresses on ae and fc due to the uniform load. The stress on fc—or that which tends to produce a stress on fc—is that due to the stress on a, and that on ae that due to the load on c, and also on all the joints between c and B; but as the latter is constant, we have occasion now to consider only the load wx on ac.

Of the load on a, B sustains $\dfrac{n-1}{N} \times \tfrac{1}{2}\dfrac{wx^2}{l}$, and of the load on c, A sustains $\dfrac{N-n}{N} \times wx\,\dfrac{(l-\tfrac{1}{2}x)}{l}$.

Hence, the stress on ae due to both is the difference of these multiplied by $\sec \theta$, or

$$\left[(n-1)x^2 - (N-n)(2l-x)x\right]\frac{w \sec \theta}{2Nl},$$

which is to be a maximum.

$$\therefore 2(n-1)x - 2(N-n)(l-x) = 0$$
$$\therefore x = \frac{N-n}{N-1}l,$$

and the whole length of the load is $bB = x + (N-n)l = (N-n)\dfrac{Nl}{N-1}$.

The fraction $\dfrac{Nl}{N-1}$ is constant, and its value is equal to the span divided into one less number of divisions than there are bays in the chord. Hence, to produce a maximum stress for an uniform load, conceive that the span is divided into one less number of spaces than there are bays, and that for the successive braces the maximum stress is produced when the uniform load extends from the division under the brace to the end towards which it inclines. The maximum thus produced, including the dead and live loads, is

$$F_n = \left\{\begin{array}{l}\dfrac{(N-n)(N-n-1)}{2N}wl + \tfrac{1}{2}wl \times \dfrac{N-n}{N} \\ + \dfrac{(N-1)}{2Nl}\left(\dfrac{N-n}{N-1}l\right)^2 w - \dfrac{N-n}{N}\left(\dfrac{N-n}{N-1}\right)lw \\ + \tfrac{1}{2}(N - 2n + 1)w\end{array}\right\} \sec \theta \quad (133)$$

which is less than that produced by equal weights placed at the joints; and hence the formulas previously developed are on the safe side.

113.—STRESS ON THE VERTICALS.— According to principle c, article 71, if the load is on the supported chord the stress on the n-th vertical is the same as the vertical component of stress on the n-th diagonal. The same may be seen by examining Fig. 80. It must be observed, however, that unnecessary members are not included in the count. In the case of the *Howe type*, the outline of the truss is trapezoidal in form, as in Fig. 77, although practically it is usually made rectangular, somewhat like Fig. 79. In the *Pratt type*, the truss is usually rectangular, as in Fig. 79, but if it is supported at the upper chord, it may be trapezoidal, like Fig. 77, inverted. In either or any of these cases, principle e of article 71 is applicable. Hence if the load is upon the supported chord we have only to make $\sec \theta = 1$ in Eq. (128) for the maximum stress on the n-th vertical. Hence

The maximum stress on the n-th vertical of the panel system for equal weights placed on the supported chord is

$$\left[(N-n)(N-n+1)p + (N-2n+1)Nw,\right]\frac{1}{2N} \quad \ldots (134)$$

in which n must not exceed $\frac{1}{2} N$.

If the load be upon the upper chord, Fig. 80, the maximum stress on the verticals may be found by subtracting $p + w_1$ from the preceding expression. Or we may observe that the stress on the n-th vertical is the same as the vertical component of the stress on the $(n + 1)^{th}$ diagonal brace.

Hence in the "Howe type," with equal weights on the unsupported chord, the maximum stress on the $(n-1)^{th}$ vertical is given by Equation (134), by substituting the proper value for n. Thus, for the 5th vertical make $n = 6$.

If the load be upon the upper chord, in Fig. 79, the stress on the n-th vertical is the same as the vertical component of the stress on the $(n-1)^{th}$ diagonal tie. Hence, in this case Eq. (134) gives the stress on the $(n + 1)^{th}$ vertical.

128   TREATISE ON BRIDGES.

NOTE 1.—This reasoning appears direct, simple, and correct, but it is not difficult to raise doubts as to its accuracy in the minds of some.* Thus, in Fig. 80, if the joints from 2 to g are loaded, the stress on Aa, $a1$, $1b$, and $b2$, has been found to be $V_1 \sec \theta$; and the value of V_1 is the sum of all the components of the several weights which are sustained at A. For instance, $\frac{3}{5} p_2$ is supported at A. But the whole of p_2 is transmitted through $2b$ to b. Should we not therefore add to the value found by Eq. (134) $\frac{3}{5} p_2$ or generally $\frac{n}{N} p$? I say that it would be incorrect to add $\frac{n}{N} p$. The formula is correct without it, for it may be shown that the formula recognizes the fact that all of p_2 is transmitted through $b2$. I observe, 1st, that if p_2 were the only load, then it would be necessary to add $\frac{3}{5} p_2 \sec \theta$ to the reaction $V_1 \sec \theta$; or to $\frac{3}{5} p_2 \sec \theta$; but in this case $b3$ would be strained as well as $b1$. The *coefficient* due to p_2 on $b3$ is 2. I observe. 2d, that p_4 produces a stress on $c2$ of 2; and hence the resultant stress on $b3$ and $c2$ is zero; and so far as these two weights are concerned they might be removed. The resultant stress on the vertical is $\frac{3}{5} p + \frac{2}{5} p = \frac{5}{5} p$ $= p$, which is the value of p_3 when $p = p_3 = p_4$. So that the formula leads to the correct result. The addition of the weights p_2 and p_4 does not affect the stress just found; but these additions add to the stress on $b2$ and $c2$.

We state then, generally, that the vertical component of the main brace gives the stress on one or the other of the adjacent verticals for any distribution of the load *except when the counter-brace in the next panel is strained ; in which case the vertical component of the stress on the counter-brace must be added to the former.*

It is evident that both the *main* and *counter* in the same panel cannot be strained at the same time by the action of a load.

The reasoning here given may be made more evident by using Fig. 49. If we there add $\frac{1}{2}p$ to the result given by the formula for the stress on EC, we shall have $\frac{1}{2}p$, for the p at C, $+\frac{1}{2}p$, for the p at D (which together make p, which is the amount carried at A), $+\frac{1}{2}p$ at C for the amount carried at B; making $\frac{1}{2}p$ to be carried by EC. But it is evident that EC and FD both carry only $2p$ and hence each carries p. This shows the absurdity of adding the fractional part of p which is carried by the opposite support when the counter diagonal is not strained.

NOTE 2.—The student may think that in practical cases confusion may arise in distinguishing between the Pratt and Howe types, and the position of the loading whether it be upon the supported or unsupported chords. But computations in such cases are made deliberately, and suitable checks are put upon the work; and if there be any doubt in regard to the proper formula, it is better to consider the effect of each weight separately upon each piece of the truss. and tabulate the results, as on page 110. In the case of bridges —especially rail-road bridges—the load is rarely composed of *equal* weights at the joints; for the locomotive is often two or three times as heavy as the same length of any portion of the remainder of the train, in which case the formulas previously given are not applicable. They may be used by applying

---

* See *Trautwine's Eng. Pocket-Book*, p. 275, including foot-note. Also *Am. R. R. Times* April 17, 1869.

them to all that portion of the load which is uniform, and then adding the effect of the extra load or loads which are applied at the joint or joints.

NOTE 3.—As a check upon the work the problem may be solved geometrically and the results determined by a scale. There are several ways of doing this. One is by the method used in Fig 60. Another is by the method of diagrams as developed by Rankine and Clerk Maxwell, and which is illustrated in Figs. 40 and 41. The latter method when well understood is easily applied and gives satisfactory results. Geometrical methods are exceedingly valuable for securing general results, and checking analytical work. They are more reliable for general results than analytical methods, for large errors are less likely to happen; but they are not as precise, for they do not give the nearest fraction as certainly as numerical computations do.

114.—STRESS UPON THE CHORDS.—The maximum stress upon all parts of the chords, for equal weights at the joints, exists when all the joints are loaded. Remembering that for statical equilibrium the algebraic sum of the moments of all the external forces will be zero, and we readily find the reaction V_1 of the support, Fig. 80, by taking the moments of all the weights and the reaction, about the point g. To find the internal stress, we may conceive that one of the chords is severed ; as, for instance, the bay 3—4, Fig. 80, and suppose that an external force is substituted for the stress. Then take the origin of moments at the point about which the system would turn if the stress-force were removed.* Thus, if the bay 3—4 be severed, the system will evidently tend to turn about c. It will generally tend to turn about that joint which is nearer the centre, for a uniform load. In this case if $d3$ were in action, instead of $c4$, it would tend to turn about d.

Let t_n = the stress on the n-th bay of the lower chord (as 3–4, Fig. 80);

c_n = the stress on the n-th bay of the upper chord (as cd);

D = the depth of the truss;

and the other notation as previously used.

For stress on the lower chord, we have the following equation of moments:

$$V_1 n\, l - p_1(n-1)\, l - p_2(n-2)\, l - \text{etc., to } n \text{ terms}\ldots$$
$$= t_n D \ldots\ldots\ldots\ldots\ldots\ldots\ldots\ldots\ldots\ldots\ldots\ldots(135)$$

---

* It is not necessary to take the origin of moments at this point, but by so doing the solution is simplified, because it eliminates some of the moments which might otherwise appear.

and if $p_1 = p_2 = p_3 =$ etc., we have
$$V_1 = \tfrac{1}{2}(N-1)p, \text{ and}$$
$$\tfrac{1}{2}(N-1)nlp - \tfrac{1}{2}n(n-1)pl = t_n D,$$
and if the weight of the truss be included, as it should be, we finally have, by reduction, after substituting $(p + w_1)$ for p,

$$t_n = (N - n)\, n\, (p + w_1) \frac{l}{2\,D} \quad \ldots \ldots \ldots \ldots \ldots (136)$$

in which n must not exceed $\tfrac{1}{2} N$ for N even, or $\tfrac{1}{2}(N+1)$ for N odd. (If, however, the two central bays 2–3 and 3–4 be considered as one, the formula will give the stress on the bays from $n = 1$ to $n = (N-1)$.)

Stress on the upper chord.—If we call ab (Fig. 80) the first bay of the upper chord, bc the second, and so on, and calling bc the n-th, and taking the origin of moments at 2, we find the same result as in the preceding equation.

It is well to observe that instead of dealing with the series given above for the weights p_1, p_2, etc., we may find the total value of their moment by multiplying their sum by the distance of their centre of action from the origin of moments.

An analysis of the Pratt System, Fig. 79, gives exactly the same formula, if in the lower chord we neglect $A\,1$ in the count for the number of the bay. Thus, 1–2 would be called the first bay.

It makes no difference in any of these cases whether the load be upon the upper or lower chord.

Hence, in the panel system for a uniform load throughout, the stress on the chords varies as the product of the segments into which the span is divided by the joint about which the frame would turn if the bay were severed. Fig. 81 is an illustration of this law.

If $n = \tfrac{1}{2} N$ we have for the stress at the middle of the span

$$\tfrac{1}{8}\frac{WL}{D} \quad \ldots \ldots \ldots \ldots \ldots \ldots \ldots \ldots \ldots \ldots (137)$$

115.—EXAMPLES.—1. Let $p = w_1 = 5$ tons (net), $N = 12$, $\theta = 45°$ ∴ $l = D$. Required the stress on each bay of the lower chord, Howe Type.
From Eq. (136) we readily find:—

$t_1 = 55$ tons; $t_2 = 100$; $t_3 = 135$ · $t_4 = 160$; $t_5 = 175$; $t_6 = 180$
First diff.   45   35   25   15   5
Second diff.   10   10   10   10

PANEL SYSTEM.   131

2. Required the depth so that the stress at the middle shall equal one-half the total load. Use Equation (137).

3. Required the depth so that the stress at the middle shall equal $\frac{1}{n\text{-}th}$ of the total load.

4. If $l = D$, and $p = w_1$; and for a uniform load the stress on the second bay is 32 tons; and on the fourth bay it is 48 tons. Required the number of bays and the value of the uniform load.

5. For uniform live and dead loads, the stress on the first bay is $22\frac{1}{4}$ tons; on the second, 42 tons; on the fourth, 72 tons; on the sixth, 90 tons; on the eighth, 96 tons. Required the total number of bays, the weight of the live and dead loads, the depth of the truss, and the length of one bay.

Ans. $N = 16$, $p = 2$ tons, $w_1 = 1$ ton, and $l = D$.

116.—LOAD CONCENTRATED AT ONE JOINT.—If a load P is concentrated at one joint; as, for instance, at the end of the n-th bay, and the weight of the truss be also considered; we have, if we count n from the end at which V_1 is used:

$$V_1 = \tfrac{1}{2}(N-1)w_1 + \frac{N-n}{N}P$$

and the shearing stress on the x-th diagonal is

$$S_s = V_1 - (x-1)w_1 = \tfrac{1}{2}(N-2x+1)w_1 + \frac{N-n}{N}P$$

and hence the stress on the x-th diagonal, for $x < n$

$$F_x = \left[\tfrac{1}{2}(N-2x+1)w_1 + \frac{N-n}{N}P\right]sec\,\theta.(138)$$

and for $x > n$ it is

$$F_x = \left[\tfrac{1}{2}(N-2x+1)w_1 - \frac{n}{N}P\right]sec\,\theta\ldots(139)$$

These Equations may be discussed in the same way as Equation (128), but there is not much advantage in doing so, since we rarely have to deal with the effect of a single weight. Bridges, in practical cases, are loaded at several points at the same time, and it is always on the safe side to assume that they are loaded throughout. Still there are cases, especially in long spans, in which we may err unnecessarily on the safe side by assuming a load equal, for instance, to a train of locomotives. It may be advisable to consider a train as a uniform load equal to that of a train of loaded freight cars, and then add the excess of the weight of the locomotive over that of the freight cars

This being the case, we should have a uniform dead load, a partial uniform live load, and a concentrated load, in which case the maximum stress on the n-th diagonal would be

$$\left[(N-x)(N-x+1)p + (N-2n+1)Nw_1 + 2(N-n)P\right]\frac{\sec\theta}{2N} \ldots\ldots\ldots\ldots(140)$$

But it may be better in such cases as that just considered to solve the problem numerically, for each position of the load, instead of relying upon a formula and using the formula as a check upon the work.

EXAMPLES.—1. A bridge 36 feet long; each bay 3 feet; depth 4 feet, weight 1800 lbs. A weight of 200 lbs. is placed on the 9th vertical tie. Required the strain on each of the braces.

Here $N = 12$; $\sec\theta = 1.25$; $w_1 = 150$ lbs.; $P = 200$;

$n = 9$, which substituted in Eq. (138) gives:—Strain on 1st brace, $1093\frac{3}{4}$ lbs.; 2d, $906\frac{1}{4}$ lbs.; 3d, $718\frac{3}{4}$ lbs.; 4th, $531\frac{1}{4}$ lbs.; 5th, $343\frac{3}{4}$ lbs.; 6th, $156\frac{1}{4}$ lbs.; 7th, $-31\frac{1}{4}$ lbs.; hence it is $+31\frac{1}{4}$ lbs. on the brace inclined the other way from the preceding. As there are six main braces, it appears that for this case no counter-braces are necessary. 8th, $-218\frac{3}{4}$ lbs.; 9th, $-406\frac{1}{4}$. For the main braces beyond the 9th, use Eq. (139), or we may use Eq. (138) by counting from the other end, in which case $n = 3$, and $x = 1, 2, 3$. In either case we find strain on 10th, $843\frac{3}{4}$ lbs.; 11th, $1031\frac{1}{4}$ lbs.; 12th, $1218\frac{3}{4}$ lbs. It will be seen that this series has the common difference $187\frac{1}{2}$, and the difference of the stresses on braces equi-distant from the ends is 125 lbs.

2. If $N = 10$; $n = 8$; $P = 8\,w_1$; where is the vertical force zero?
Ans. $n_0 = 7.1$.

3. If $N = 10$; $P = 8\,w_1$, where must P be applied so that the vertical force at *that point* shall be zero? Ans. $n = 7\frac{1}{4}$.

4. If $N = 8$, required the greatest concentrated load which is allowable when there are no counter-braces. Here $n = 5$; $N = 8$. $\therefore P = 1\frac{1}{4}\,w_1$.

Suppose there is one counter-brace; then $n = 6$, and we find $P = 6\,w_1$.

In a similar way we may proceed when there are two or more concentrated loads. We would find that for two concentrated loads each equal P, placed equi-distant from the ends, that the strain on the braces between the loads would be due to the weight of the bridge only.

117.—STRESS ON THE CHORDS FOR AN UNEQUALLY DISTRIBUTED LOAD.—When the load is unequally distributed the maximum stress may not be at the middle of the span, but the actual stress on any bay may be found by means of Equation (135).

MINIMUM MATERIAL IN THE PANEL SYSTEM. 133

118.—MULTIPLE SYSTEM.—If in the simple system, for a given depth of truss and inclination of diagonals the bays would be too long, they may be reduced in length by causing the diagonals to cross one or two verticals, as in Figs. 89 and 92. Such cases may be analyzed, approximately, as in article 97, or more exactly by supposing that the truss is divided into two or more simple trusses, as in article 99.

MINIMUM AMOUNT OF MATERIAL.

It is hardly possible, by a direct solution, to find all the conditions for a minimum amount of material in a truss which is composed of many parts, and all of whose parts are correctly proportioned to resist the stresses to which they are subjected. The reason of this will appear hereafter. We may, however, solve some of the more simple cases, and from them derive some important hints.

119.—PROBLEM.—GIVEN THE SPAN AND DEPTH OF A PANEL TRUSS, IT IS REQUIRED TO FIND THE INCLINATION OF THE DIAGONALS SO AS TO SECURE A MINIMUM AMOUNT OF MATERIAL IN THE DIAGONALS, THE VERTICALS AND CHORDS BEING OF UNIFORM SIZE THROUGHOUT, AND THE LOAD BEING UNIFORM THROUGHOUT THE WHOLE LENGTH.

In the Pratt Truss the diagonals are ties, and hence their sections should be directly proportional to the stress to which they are subjected; but in the Howe Truss, the diagonals being braces, their section, if they are long compared with their diameter, will be a function of their length as well as of the stress: see article 21. This fact complicates the problem. *For the sake of simplicity in this problem, we will therefore assume that the section of the diagonals is directly proportional to the stress.*

Referring, therefore, to Fig. 77 or Fig. 79, we

let $L = AB =$ the span;
$D =$ the depth of the truss;
$N =$ the number of bays;
$l = L \div N =$ the length of a bay;

W_1 = the weight of the frame;
$w_1 = W_1 \div N$ = the weight at one joint of the dead load;
W_2 = the weight of the live load;
$p = W_2 \div N$ = the weight on one bay of the live load;
$W = W_1 + W_2$ = the total load;
$r = \dfrac{w_1}{p}$;
$\theta = a\, 1\, b$ = the inclination of the diagonal to the vertical;
k = the transverse section of a diagonal; and,
T = the modulus of strain, that is, the strain per unit of transverse section;

Then $p + w_1 = (1 + r)\dfrac{W_2}{N}$;

$$D = \dfrac{l}{tang\,\theta} = \dfrac{L}{N\,tang\,\theta}$$

The length of a brace $= D\sec\theta = \dfrac{L\sec\theta}{Nt\,ang\,\theta}$

The stress on the n-th brace is, see Eq. (129a).

$\tfrac{1}{2}(N - 2n + 1)(p + w_1)\sec\theta = Tk$

$\therefore k = (N - 2n + 1)\dfrac{(p + w_1)\sec\theta}{2\,T}$ ......(141)

Making $n = 1, 2, 3$, etc., to $\tfrac{1}{2} N$ (observing that for $n = \tfrac{1}{2}(N + 1)$ the expression becomes zero, and we have

$k_1 = (N - 1).\dfrac{(p + w_1)\sec\theta}{2\,T}$

$k_2 = (N - 3).\dfrac{(p + w_1)\sec\theta}{2\,T}$

$k_{\frac{1}{2}N} = 1.\dfrac{(p + w_1)\sec\theta}{2\,T}$

Summing this series gives

$$\Sigma K = \dfrac{N^2\,(p + w_1)\sec\theta}{8\,T}\ldots\ldots\ldots\ldots(142)$$

which, multiplied by the length of a brace, gives for the *volume, of half the diagonals in all the trusses,*

$$\dfrac{N\,(p + w_1)\,L}{8\,T}\cdot\dfrac{1}{\cos\theta\sin\theta}\ldots\ldots\ldots\ldots(143)$$

which is a minimum for $\theta = 45°$. Hence the minimum amount of material in the diagonals for both trusses is

$$\frac{N(p+w_1)L}{2\,T} = \frac{W_2+W_1}{2\,T}L = \tfrac{1}{2}\,W\frac{L}{T}\ldots(144)$$

which, compared with Eq. (l) article 38, shows that the material in the diagonals is the same as that in the rafters of a king-post truss having the same span and same inclination. Observe that P Eq. (l) above referred to equals $\tfrac{1}{2}W$ in Eq. (144), and $T = C$.

NOTE.—This result may at first appear to be incorrect, since the rafters of the king-post are of uniform size, while in the panel system the diagonals diminish in size from the ends toward the centre. But it should be observed that in the king-post both rafters carry only one-half the total uniform load; while the end diagonals of the panel system carry nearly the total uniform; or exactly $(N-1)(p+w_1)$. One-half the load in both cases on the end bays is carried directly by the supports; but in the king-post this distance is one-half the span, while in the latter it may be only a small fraction of the span.

We may now carry the solution further. If the verticals are of uniform size, they must be of the size as the first one. Hence, in Eq. (141) make $sec.\ \theta = 1$ and $n = 1$ and we find for the section of each of the verticals

$$\tfrac{1}{2}(N-1)\frac{(p+w_1)}{T}$$

Hence, their volume $= \tfrac{1}{2}(N-1)^2 D\dfrac{p+w_1}{T} = \tfrac{1}{2}(N-1)\dfrac{L}{N}\dfrac{(p+w_1)}{T}$

which expression diminishes as N diminishes; hence the deeper the truss the less the amount of material will be required for the *diagonals and verticals*. The king-post with rafters inclined at an angle of $45°$ requires least.

The section of the chords will be, see Eq. (137)

$$K = \frac{W\,L}{8\,D\,T} = \frac{W\,N\,l}{8\,D\,T} = \frac{W\,N}{8\,T},\ \text{if}\ l = D,$$

from which we see that K varies directly as N, and hence is least when N is least, or equal to 2. In this case the upper chord disappears. Hence, considering all the parts of the truss, the king-post with rafters inclined 45 degrees requires less material than the panel system.

NOTE.—It does not follow from this that the king-post should in all cases be substituted for the panel system; for in most cases it would be quite impracticable to do so, on account of the great depth which would result. For instance, in a 100-foot span the depth would be 50 feet!

120.—A SECOND PROBLEM OF MINIMUM MATERIAL.—NEXT SUPPOSE THAT THE DIAGONALS AND VERTICALS ARE BOTH PROPORTIONED FOR THE STRAINS DUE TO A UNIFORM LOAD, ASSUMING, AS BEFORE, THAT THE SECTIONS OF THE PIECES ARE PROPORTIONAL TO THE STRESSES.

The stress on a vertical is (Eq. (141)),

$$\tfrac{1}{2}(N - 2n + 1)(p + w_1)$$

hence its volume is $\tfrac{1}{2}(N - 2n + 1)\dfrac{(p + w_1)L}{T N \tan \theta}$.

Making $n = 1, 2, 3$, etc., to $\tfrac{1}{2} N$, and summing the series, and we find that the volume of all the verticals is

$$\frac{N(p + w_1)L}{8 T} \cdot \frac{1}{\tan \theta}$$

which added to Eq. (143) gives for the total volume of all the diagonals and verticals

$$\frac{N(p + w_1)L}{8 T}\left(\frac{1}{\sin \theta \cos \theta} + \frac{1}{\tan \theta}\right)\ldots\ldots(145)$$

which is a minimum for $\theta = 54° 45'$.

NOTE.—In practice the diagonals in wooden trusses are generally inclined less than 45 degrees from the vertical.

121.—GENERAL PROBLEM OF MINIMUM MATERIAL.—LET THERE BE A PERMANENT DEAD LOAD AND A MOVING LIVE LOAD; IT IS REQUIRED TO FIND THE CONDITIONS FOR A MINIMUM AMOUNT OF MATERIAL IN THE DIAGONALS, VERTICALS, AND CHORDS, THE TRANSVERSE SECTION OF ALL THESE PARTS BEING PROPORTIONED DIRECTLY AS THE STRESSES TO WHICH THEY ARE SUBJECTED.[*]

Volume of the diagonals.

We find the section of the n-th diagonal by means of Eq. (128) to be

---

[*] This solution was given by William Bouton, class of 1865, *Univ. of Mich.* See *Jour. Frank. Inst.*, Vol. L., 3d series, p. 76.

MINIMUM MATERIAL IN THE PANEL SYSTEM.   137

$$K_n = \left[(N-n)(N-n+1)p + (N-2n+1)Nw_1 \right] \frac{\sec \theta}{2\,NT}$$

which, multiplied by the length $(L \div N) \div \sin \theta$ gives for the volume of the n-th diagonal

$$Vol._n = \frac{\left[(N-n)(N-n+1)p + (N-2n+1)Nw_1 \right] L}{2\,N\,T \sin \theta \cos \theta} \quad \ldots \ldots (146)$$

The number of diagonals which incline either way is N_o, see article 106; hence to find the volume of one-half the diagonals, make $n = 1, 2, 3$, etc., to N_o in Eq. (146), and take the sum of the results. Two series will thus be formed, one the coefficients of p; the other of w_1. The sum of the series may readily be found by the method of differences. The sum will be

$$S = n'a + \frac{n'(n'-1)}{2} d_1 + \frac{n'(n'-1)(n'-2)}{2.3} d_2 + \text{etc}\ldots(147)$$

in which $S =$ the sum;
$\qquad a =$ the first term of the series;
$\qquad d_1, d_2$ etc., $=$ the first term of the successive orders of differences; and
$\qquad n' =$ the number of terms in the series $= N_o$.

First consider the coefficient of p.
Making $n = 1, 2$, etc., in Eq. (146), and we have,

the series, $N(N-1)$, $(N-1)(N-2)$, $(N-2)(N-3)$, $(N-3)(N-4)$, etc.

1st Dif. $\quad -2N+2 \quad\quad -2N+4 \quad\quad -2N+6$
2d Dif. $\quad\quad\quad +2 \quad\quad\quad\quad +2 \quad\quad\quad\quad +2$
3d Dif. $\quad\quad\quad\quad\quad 0 \quad\quad\quad\quad\quad 0$

Hence, $n' = N_o$, $a = N(N-1)$, $d_1 = 2(-N+1)$ and $d_2 = 2$.

$$\therefore S = \left[N_o N(N-N_o) + \tfrac{1}{3} N_o (N_o^2 - 1) \right].$$

Second, the coefficient of w_1.

This becomes negative when we pass the centre, and each negative term will cancel an equal positive one; hence we need sum only $N - N_o$ terms.

The series is $N(N-1)$, $N(N-3)$, $N(N-5)$, $N(N-7)$
1st Dif. $-2N$ $-2N$ $-2N$
2d Dif. 0 0

$$\therefore S = N_o N(N - N_o)$$

Adding these results and multiplying by the common factor $\dfrac{L}{2 N^2 T \sin \theta \cos \theta}$ gives for the volume of half the braces =

$$\dfrac{\left[N N_o (N - N_o)(p + w_1) + \tfrac{1}{3} N_o (N_o^2 - 1) p \right] L}{2 N^2 T \sin \theta \cos \theta} \dots \dots (148)$$

Volume of the Verticals.

The stress is, $\left[(N - n)(N - n + 1) p + (N - 2n + 1) N w_1 \right] \dfrac{1}{2N}$.

The length is, $D = \dfrac{L}{N \tan \theta}$

$\therefore$ Volume of n-th vertical $= \left[(N - n)(N - n + 1) p + (N - 2n + 1) N w_1 \right] \dfrac{L}{2 N^2 T \tan \theta} \dots \dots (149)$

Summing the series as before from $n = 1$ to $n = \tfrac{1}{2}(N - 1)$ (for N odd), and we have for the volume of half the *verticals*.

$$\dfrac{\left[\tfrac{1}{4} N(N^2 - 1)(p + w_1) + \tfrac{1}{6}(N - 1)(\tfrac{1}{4}(N-1)^2 - 1) p \right] L}{2 N^2 T \tan \theta} \dots \dots (150)$$

If N is even, we sum the series from $n = 1$ to $n = \tfrac{1}{2} N$, and deduct one-half the volume of the middle vertical, the value of which may be found from Eq. (149). This done and we find for half the volume of the verticals when N *is even*

$$\left[\tfrac{1}{4} N^2 (p + w_1) + \tfrac{1}{24} N [N(N-3) - 10] p \right] \frac{L}{2 N^2 T \tan \theta} \quad (151)$$

Volume of the Chords.

The stress on the n-th bay of the upper or lower chords is

$$\tfrac{1}{2} (N - n) n (p + w_1) \tan \theta \dots \dots \dots (152)$$

The limits of n are from $n = 1$ to $n = \tfrac{1}{2} N$ for the lower chord, and from $n = 1$ to $n = \tfrac{1}{2} (N - 2)$ for the upper chord when N is *even*; and when N is *odd* the limits for both chords are $n = 1$ to $n = \tfrac{1}{2} (N - 1)$ by omitting half the central bay in the lower chord and including it in the upper, as we may do since the stress is the same in both. In either case, the sum of the series found by making $n = 1, 2, 3$, etc., in (152) multiplied by $l = \dfrac{L}{N}$ and otherwise reduced is

$$\tfrac{1}{6} N (N - 1) (p + w_1) \frac{L}{2 T N} \tan \theta \dots \dots \dots (153)$$

Hence, the total volume of half the truss is the sum of (148), (150) or (151), as N is odd or even, and (153), or

$$\left\{ \begin{array}{l} [N N_\bullet (N - N_\bullet) (p + w_1) + \tfrac{1}{3} N_\bullet (N_\bullet^2 - 1) p] \frac{1}{\sin \theta \cos \theta} \\ + \begin{cases} \text{(for } N \text{ odd)} \tfrac{1}{4} N (N^2 - 1)(p + w_1) + 1 \cdot 6 (N - 1)(\tfrac{1}{4}(N-1)^2 - 1; p \\ \text{(for } N \text{ even)} \tfrac{1}{4} N^2 (p + w_1) + 1 \cdot 24 N [N(N - 3) - 10] p \end{cases} \frac{1}{\tan \theta} \\ + 1 \cdot 6 N^2 (N - 1) (p + w_1) \tan \theta \end{array} \right\} \times \frac{L}{2 N^2 T} (154)$$

which is to be a minimum.

$$\cdot \left\{ \begin{array}{l} [N N_\bullet (N - N_\bullet) (p + w_1) + \tfrac{1}{3} N_\bullet (N_\bullet^2 - 1) p] (\tan^2 \theta - 1) \\ - \begin{cases} \text{(for } N \text{ odd)} \tfrac{1}{4} N (N^2 - 1)(p + w_1) + 1 \cdot 6 (N - 1)(\tfrac{1}{4}(N-1)^2 - 1) p \\ \text{(for } N \text{ even)} \tfrac{1}{4} N^2 (p + w_1) + 1 \cdot 24 N [N(N - 3) - 10] p \end{cases} \\ + 1 \cdot 6 N^2 (N^2 - 1) (p + w_1) \tan^2 \theta \end{array} \right\} = 0 \dots (155)$$

from which $\tan \theta$ may be directly found.

EXAMPLE.—Let $p = w_1$ and $N = 10$. Then, according to the table in article 108, $N_\bullet = 6$. These values in Eq. (155) give $\tan^2 \theta = 0.27916 \therefore \theta = 27° 50'$.

We shall also have $l = 0.528 \, D$ and $L = 10 \, l = 5.28 \, D \therefore L + D = 5.28$. In this way the following table is formed.

TABLE

Showing the inclination of the diagonals and the ratio of the length to the depth in the Panel System for a Minimum Amount of Material, the load being on the lower chord.

Values of N.	$p = 0$.		$p = w_1$		$p = 2w_1$		$w_1 = 0$ p	
	θ	$\dfrac{L}{D}$	θ	$\dfrac{L}{D}$	θ	$\dfrac{L}{D}$	θ	$\dfrac{L}{D}$
3.	39° 13′	2,448	39° 35′	2,481	39° 42′	2,490	39° 54′	2,508
4.	37° 04′	3,020	37° 03′	3,020	37° 01′	3,016	37° 21′	3,052
5.	34° 11′	3,395	34° 50′	3,480	35° 02′	3,505	35° 38′	3,585
6.	32° 35′	3,834	32° 42′	3,852	33° 13′	3,930	33° 49′	4,020
7.	30° 43′	4,158	31° 29′	4,284	31° 40′	4,319	32° 13′	4,410
8.	29° 28′	4,520	30° 04′	4,632	30° 12′	4,656	31° 05′	4,824
9.	28° 06′	4,806	28° 51′	4,959	29° 10′	5,022	29° 59′	5,193
10.	27° 10′	5,130	27° 50′	5,230	28° 06′	5,340	28° 54′	5,520
12.	25° 16′	5,664	25° 24′	5,700	26° 17′	5,928	27° 07′	6,144
15.	23° 05′	6,390	23° 51′	6,630	24° 08′	6,720	24° 59′	6,840
20.	20° 30′	7,480	21° 12′	7,760	21° 30′	7,850	22° 18′	8,200
30.	17° 07′	9,240	17° 38′	9,540	18° 03′	9,780	18° 50′	10,230
40.	15° 16′	10,920	15° 39′	11,200	15° 51′	11,560	16° 36′	11,920
50.	13° 33′	12,050	14° 25′	12,450	14° 25′	12,850	14° 25′	12,950

122.—MINIMUM MATERIAL IN A POST AND TIE *for carrying a given weight, when the distance between their lower ends is constant, and the distance from the vertex to the lower chord is given.*

In the transmission of strains we have seen that the vertical component of stress is constant between two loaded joints. Let abc be one of the triangles of a truss, ab the bay, bc the tie, and ac the post, and no load at c.

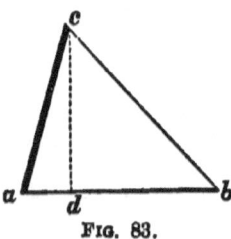

Fig. 83.

Let $l = ab =$ the length of a bay;
$l_1 = ac =$ the length of the post;
$D = cd =$ the depth of the truss;
$P =$ the vertical stress at b which is to be transmitted to a, and so on, to the support;
$x = ad$; $y = db$, and
$\delta =$ the weight of a unit of volume of the tie.

The post ac is supposed to be so long that it is in danger of bending. Suppose that the ends are rounded and that it is

MINIMUM MATERIAL IN THE PANEL SYSTEM. 141

made of solid cast iron. Then, according to article 25, its weight will be

$$0.023926469 \ (P' \ l_1^{1.88})^{\frac{1}{1.88}} = c \ (P' \ l_1^{1.88})^{\frac{1}{1.88}}$$

in which P' is the pressure in the direction of the length of the post, and equals $P \sec. acd = P \dfrac{\sqrt{D^2 + x^2}}{D}$ and c is the constant coefficient. The length of the post will be $l_1 = \sqrt{D^2 + x^2} \therefore l_1^{1.88} = (D^2 + x^2)^{\frac{3.58}{2}}$ Hence, the weight will be

$$c \left(\frac{P \ (D^2 + x^2)^{1.29}}{D} \right)^{\frac{1}{1.88}}$$

The stress on cb will be $P \sec. dcb$; its section, $K = \dfrac{P \sec. dcb}{T}$;

its volume $= \dfrac{P \sec. dcb}{T} \times D \sec. dcb$; and its weight $=$

$\dfrac{\delta \ PD \sec.^2 dcb}{T} = \dfrac{\delta \ P \ (D^2 + y^2)}{T.D} = \dfrac{\delta \ P \ (D^2 + (l-x)^2)}{T.D}$ Hence

the total weight of both pieces will be

$$c \left(\frac{P \ (D^2 + x^2)^{1.29}}{D} \right)^{\frac{1}{1.88}} + \frac{\delta \ P \ (D^2 + (l-x)^2)}{TD} \ldots \ldots (156)$$

which is to be a minimum. It may be put under the form

$$\frac{c.T.D^{\frac{1.79}{1.88}}}{\delta \ P^{\frac{.88}{1.88}}} (1 + \tan^2 \theta)^{\frac{2.29}{1.88}} + \left(\frac{l}{D} - \tan \theta \right)^2 + 1 \ldots (157)$$

in which $\tan \theta = \dfrac{x}{D}$.

Differentiating Eq. (156) and placing it equal to zero, and it gives

$$\tfrac{2.29}{1.84} c \left(\frac{P}{D} \right)^{\frac{1}{1.88}} (D^2 + x^2)^{\frac{.41}{1.88}} x + \frac{2 \ \delta \ P}{T.D} x - \frac{2 \ \delta \ Pl}{T.D} = 0.$$

This cannot be solved in finite terms. We may, however, for known values of the constants find the value of x to any degree of approximation. We may also deduce from it some general facts.

142   TREATISE ON BRIDGES.

The tangent of the inclination of the post is $\dfrac{x}{D} = tang\,\theta$

$$\therefore \tfrac{2\cdot 2\cdot 2}{0\cdot 94}c\left(\dfrac{P}{D}\right)^{1.88}(1 + tang^2\,\theta)^{\frac{41}{188}} x.D^{\frac{82}{188}} + \dfrac{2\,\delta\,P}{T}tang\,\theta = \dfrac{2\,\delta\,Pl}{TD}$$

or

$$\tfrac{2\cdot 2\cdot 2}{0\cdot 94}\dfrac{c.T.D^{1.70}_{1.88}}{P^{0.88}_{1.88}}(1 + tang^2\,\theta)^{\frac{41}{188}} tang\,\theta + 2\,\delta\,D\,tang\,\theta = 2\,\delta\,l \quad (158)$$

It appears from this:—

1st, That the inclination of the post for minimum material (and hence of the tie), is dependent upon the stress to be transmitted, the depth of the truss, the modulus of tenacity of the tie, the length of the bay, and the weight of a unit of the material, unless the weight per unit of the post and tie are the same. If they are the same, δ may be cancelled. As the formula now stands, δ in the first term is included in the coefficient c.

2d, No *simple relation* exists between the inclination and the other known quantities for minimum material.

3d, We know in practice that the weight of the post is much greater than that of the tie; hence we find, *approximately*, the proper inclination by neglecting all but the first term. This done and we have $tang\,\theta = 0$, and $tang^2\,\theta = -1$, or $tang\,\theta = \pm\sqrt{-1}$. The latter value being imaginary is inadmissible. The former value shows that the *posts* should be a vertical. If the posts are vertical, the other terms give $tang\,\theta = \dfrac{l}{D}$, as they should.

4th, Suppose that $tang\,\theta$ is very small. Then the term $(1 + tang^2\,\theta)^{\frac{41}{188}}$ will somewhat exceed *unity*. Suppose as a second approximation that it is *unity*, and we have

$$tang\,\theta = \dfrac{2\,\delta\,l}{\tfrac{2\cdot 2\cdot 2}{0\cdot 94}cP^{-\frac{88}{188}}D^{\frac{170}{188}}\,T + 2\,\delta\,D}$$

$$= \dfrac{188\,\delta\,l\,P^{\frac{88}{188}}}{229\,c\,TD^{\frac{170}{188}}}\text{ nearly,}$$

since $2\,\delta\,D$ will be small compared with the other term of the denominator, and may be neglected. Hence, approximately, the tangent of the inclination of the post for minimum material varies directly as the length of the bay, and nearly as the square root of the vertical stress, and inversely nearly as the depth, since the exponent of D is nearly unity.

5th, When l and D are constant, it appears that the inclination of the posts increases as P increases, and hence for minimum material should be more inclined at the ends of a truss than at the middle.

EXAMPLE.—Let $D = 18$ feet 9 inches, $l = 12$ feet 6 inches, $P = 62,000$ lbs., $s = 3$ lbs. per inch of section and one foot in length, $T = 10,000$ lbs. and $c = 0.023926469$. Required the inclination of the post for a minimum amount of material.

∴ Eq. (157) becomes

$$\frac{0.023926469 \times 10,000 \times 18.75^{\frac{170}{188}}}{3 \times 62,000^{\frac{88}{188}}} (1 + tang^2 \theta)^{\frac{212}{188}} + (\tfrac{3}{8} - tang\, \theta)^2 + 1.$$

We have log. 0.023926469 = $\overline{2}$.3788786
log. 10,000 = 4.0000000
$\tfrac{170}{188}$ log. 18.75 = 1.1404735
log. 3 ar. co. = 9.5228787
$\tfrac{88}{188}$ log. 62,000 ar. co. = 7.7567524

log. 6.42663 = 0.8079832

Hence, the expression becomes

$6.42663 (1 + tang^2 \theta)^{\frac{212}{188}} + (\tfrac{3}{8} - tang\, \theta)^2 + 1$

If $tang\, \theta = 0.00$, the expression becomes 8.09329
$tang\, \theta = 0.03$, the expression becomes 7.90246
$tang\, \theta = 0.04$, the expression becomes 7.83185
$tang\, \theta = 0.05$, the expression becomes 7.54871
$tang\ \ \ = 0.06$, the expression becomes 7.77841
$tang\, \theta = 0.08$, the expression becomes 7.82001;

from which it is evident that the minimum is for $tang\, \theta = 0.05$ nearly; or for $\theta = 2° 52'$ nearly.

SPECIAL TRUSSES OF THE PANEL SYSTEM.

123.—LONG'S TRUSS.—One panel of the Long's Truss is shown in Fig. 84. This is one of the earlier wooden bridge trusses of this country.* The upper and lower chords were

---

* See *Jour. Frank. Inst.*, Vol. V., 3d Series, p. 231, for a diagram, and claims of the patentee. See also *Am. Jour. of Arts and Science*. Vol. XXXVIII., p. 202.

composed of small stringers AAA which were spliced, when a single piece could not be obtained which would reach the whole

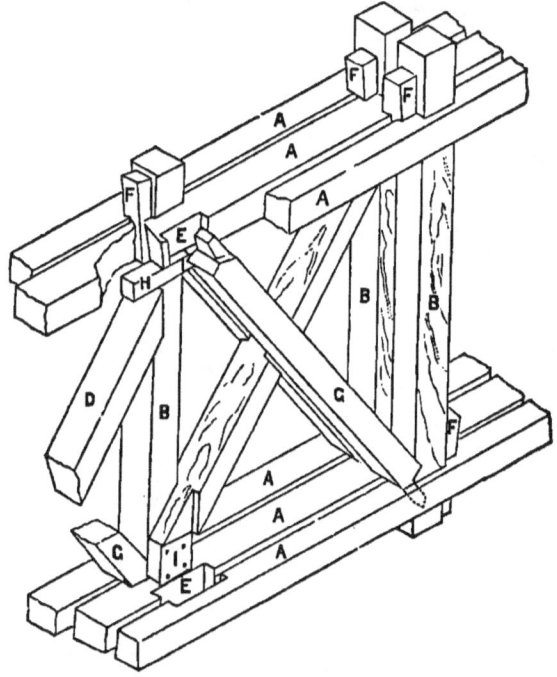

Fig. 84.

length of the bridge. The vertical posts (which served as ties) were double. The manner of securing them to the chords is sufficiently evident from the figure. The keys FFF were to force the posts against the main brace G. The main braces were also double, and so placed as to leave a small space between them, through which part of the counter-brace D might pass. The manner of securing the main braces is also sufficiently evident from the figure. The lower end of the counter-brace D, rests upon a block I, which is secured to one of the posts B, by pins or in any other convenient manner. A rest is thus formed for part of the main brace, and the remaining part bears, by means of a shoulder, directly against the post B. The upper end has a shoulder for bearing against the post, and the whole is brought into bearing by a wedge (or key) H, which is driven between the upper end of the counter-brace and the upper chord. These trusses may have served a good purpose in their day, but they are no longer in demand, since the extensive use of iron makes other forms more desirable. This truss is of the "Howe Type," and the scientific principles connected with it will be sufficiently discussed in the following articles.

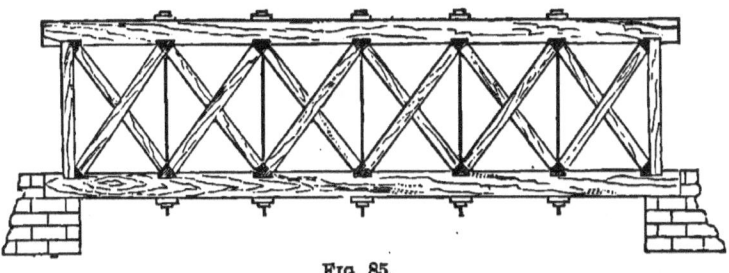

FIG. 85.

124. HOWE'S TRUSS.—The general form of Howe's Truss is shown in Fig. 85.* The diagonals are wooden braces, and verticals are iron ties. We have taken this as one of the "TYPE FORMS." As usually constructed the main braces are double, as shown in Fig. 86, and of uniform size throughout the span, and the counter-braces are of the same size, and placed between the main braces, and the chords are also of uniform size throughout the span. One of the marked features of this truss is the use of blocks which are placed between the ends of the braces and the chords. The vertical tie rods pass directly through the blocks and chords, being secured at one end by a nut, and at the other by a head, or at both ends by nuts on the rods. Such are the general features of the truss, which will now be considered more in detail.

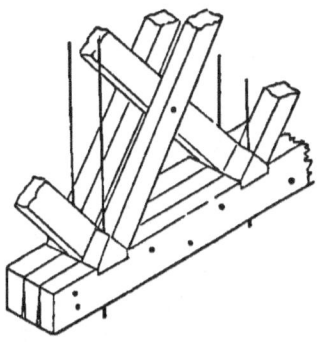

FIG. 86.

---

* *Jour. Frank. Inst.*, Vol. III., 3d series, p. 289; also *Am. Jour. of Arts and Science*, Vol. XVIII., p. 123.

125.—MAIN BRACES.—The main braces being all of the same size, their dimensions will be determined so as to resist the vertical stress at the ends. Hence if

$k =$ the section of all the braces at one end for all the trusses, and

$C =$ the modulus for crushing,

and we assume that these braces will not bend, we have directly from Eq. (129),

$$C k = \tfrac{1}{2} (N - 1) (p + w_1) \sec \theta = \tfrac{1}{2} \left[W - (p + w_1) \right]$$

$$\sec \theta = \tfrac{1}{2} W \sec \theta \text{ nearly} \ldots\ldots\ldots\ldots\ldots\ldots (159)$$

The expression $\tfrac{1}{2} W \sec \theta$ is on the safe side. But as we no longer consider the parts as reduced to rigid right lines, and the joints perfectly flexible, as was done in the preceding articles, we must consider what effect finite pieces and rigid joints will have upon the formulas.

It appears that *rigidity* cannot cause any greater stress upon the end braces than that due directly to the full loading, and if it has any effect it will diminish the stress. It seems evident that the effect of rigidity in the chords will be to diminish the stress on all the other diagonals and verticals, for if the chords were sufficiently rigid they might carry the whole load. As it is, they may carry a part of it. The transverse sections of the chords, diagonals and verticals are so small compared with the length and depth of the truss, that they can have no perceptible effect upon the formula. Equation (128) is therefore on the safe side. The longitudinal stress upon the chords is not increased in amount by the rigidity, but the tendency is to produce a little more stress on the fibres on the upper side of the upper chord and lower side of the lower chord than the average stress; and hence a somewhat less stress on the fibres on the opposite side of the chords. In short they will resist somewhat like beams. There will be a slight transverse shearing stress, depending chiefly upon the amount of deflection of the truss. On the whole, the formulas for the stress upon the chords are considered safe.

There would be a saving of timber if the braces were proportioned according to the stresses, but as timber is comparatively inexpensive, and as it would cost extra in time and labor to make the braces of different sizes, it is deemed inexpedient to do so. Indeed, in all wooden structures the tendency is to be lavish with timber rather than saving, especially when there is any question about the strength.

126.— COUNTER-BRACES.—The stress on the counter-braces is greatest at the middle of the span, and is found by making $n = \frac{1}{2}(N+1)$ in Eq. (128). Hence, the greatest stress on the counter-braces is $\dfrac{N^2-1}{8N}p \sec\theta = \frac{1}{8}Np$ nearly. This is the case when the number of bays is *odd*, in which case both the braces in the middle panel may be considered as main braces. If the number of bays is even, make $n = \frac{1}{2}N + 1$ in Eq. (128), and it reduces to $\left[\frac{1}{8}(N-1)p - \frac{1}{2}w_1\right]\sec\theta$, which is less than $\frac{1}{8}Np$. Using $\frac{1}{8}Np$ as the limit which it cannot exceed, and comparing it with Eq. (159), and we have

$$\frac{\text{maximum stress on counter-brace}}{\text{maximum stress on main brace}} = \frac{\frac{1}{8}Np}{\frac{1}{2}(N-1)(p+w_1)} = \frac{p}{4(p+w_1)} \text{ nearly} \dots\dots\dots\dots\dots\dots\dots(160)$$

The value of this fraction is always less than one-fourth; hence the counter-braces need not be one-fourth as large as the main braces. In practice they are made more than twice as large as is necessary, since the counter is usually the same size as the main, and there are half as many of them. In Eq. (160) if $w_1 = \frac{1}{4}p$, the fraction becomes $\frac{1}{5}$; if $w_1 = \frac{1}{2}p$, it becomes $\frac{1}{6}$ (and this might be considered a *practical* value); if $w_1 = p$, it becomes $\frac{1}{8}$. It has also been shown that counter-braces are not needed in all the panels, and that the effect due to the stiffness of the chords makes a still less number necessary. It would be an improvement in strength if counter-braces were placed in only a few of the central panels, see article 125, and that in all the others they were entered as main braces. Placing counter-braces in every panel, including the end ones (as is usually done in practice), seems to be without any good reason.

It is said by some that counter-braces may be used to increase the stiffness of the structure. In regard to this we observe first, that if both *main* and *counter* are strained at the same time, it is not due to the loading, but to some peculiarity in the construction, or to some artificial process. For instance, if a key is driven between the ends of a main brace and the block at the foot of it, it will tend to elongate the diagonal of the rectangle in that direction, and to shorten the diagonal in the direction of the counter-brace, and may thus induce strains on both at the same time. The same result would be secured by making both braces a little too long and forcing them into place; or by keying the *counter* instead of the *main*. The direct effect of—

127.—KEYING THE COUNTER-BRACE is to retain permanent strains upon both braces. If a uniform load is placed upon a truss, the immediate tendency will be to relieve all

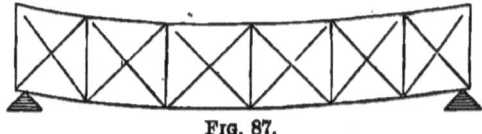

FIG. 87.

the *counters*, as in Fig. 87. If now all the *counters* are keyed, by driving a wedge between the ends of the braces and chords so as to fill the space which had been left vacant, and then the load be removed, the truss will immediately seek to regain its original form, and in doing so will bring compression upon the counter-braces, while compression still remains upon the main braces. If the counter-braces and keys were perfectly non-compressible, the same compression would exist in the main braces when the load is removed as when it is on. If this were the case, the only effect of putting the same load on as that for which it was keyed would be to relieve the strain upon the counter-braces, and there would be no increase in the deflection. But the counter-braces being compressible, they will yield somewhat under the reaction of the main braces, and a portion of the deflection which was caused by the load will be regained. As the greatest slope due to the deflection will be near the ends, where counter-braces are not needed, it appears that but little is gained in stiffness in this way, and if counter-braces are put in the end panels for the purpose of producing stiffness, they will fail of their object, unless they are so thoroughly keyed as to produce constant and heavy strains upon the other parts.

In order to key the counter-braces so as to be in bearing for a passing load, they should be keyed at the forward end of the load, as it passes on until it reaches the middle, and then the keying should be done at the rear end of the load as it passes from the middle to the end. Keying the counter-braces in this way seems to be of very doubtful utility.

128.—CAMBERING is causing the truss to be slightly arched upward. A bridge thus formed presents a better appearance to the eye than one with perfectly straight chords. It may be accomplished in various ways. Having determined the proper length of the main braces for straight chords, if their length be slightly *increased*, beginning with nothing at the centre, and increasing gradually towards the ends, any desired camber may be secured. This will give an arch-form. If they are increased in length in the opposite direction, it will tend towards a pointed form. Making the end verticals slightly shorter than the middle ones, will produce the same general effect as the former mode, for it forces the rectangle into a rhomboid with its longer diagonal in the direction of the main brace. A similar effect may be secured by making the bays of slightly different lengths.

There is no rule for determining the amount of camber, but it should evidently be such that the heaviest loads will not bring the chords to a horizontal.

129.—THE BLOCK at the ends of the braces should have its faces at right angles to the line of the brace. To accomplish this is more of a problem than at first appears. We have given

$a =$ the clear depth of a panel;
$b =$ the clear width of a panel; and
$c =$ the thickness of the brace,

to find the base and height of the block so that the hypotenuse shall equal c, and at the same time have the hypotenuse perpendicular to the line of the brace.

If $y =$ the depth of the block which rests against the side a, I find

$$4 y^4 - 4 a y^3 + (a^2 + b^2 - 4 c^2) y^2 + 2 a c^2 y = b^2 c^2 - c^4$$

which is a complete equation of the fourth degree, and its solution is impracticable. Practically the faces of the blocks may be wider than the depth of the brace, and perpendicular to the diagonal with the panel, in which case the brace may be parallel to or coincident with the diagonal.

In the cheaper structures the blocks are made of wood, but in more important ones they are made of cast iron.

The depth to which the block should be let into the chords depends upon the horizontal component of the push of the brace, and hence should be greater near the ends. No theoretical rule can be given for this case, but it is found in practice that a small depth, from one-fourth of an inch to an inch in depth, is sufficient.

To prevent the block from crushing the chord, an iron block, or hollow cone, or other suitable piece, is placed between the block which is on one side of the chord, and the washer which is on the other side, so that the vertical stress is transmitted directly to the vertical tie without pressing upon the chord. When the chords are made of several parallel pieces, this iron piece may pass between said pieces.

130.—THE VERTICAL TIE-RODS being made of iron, it will generally be more economical to proportion them according to the strains than to make them of uniform size. When they are long, it will be best to have a washer and nut at each end, instead of having a solid head at one end. The thickness of the nut should at least equal the diameter of the rod. It is a good plan to so enlarge the ends of the rods, that the diameter shall equal the body of the rod, *plus* twice the depth of the threads.

131.—THE CHORDS.—The chords being made of uniform size throughout their length, their dimensions must be determined from the stress at the middle, which is $\frac{1}{8} \frac{WL}{D}$. Of course there is an excess of timber in the chords in all but the middle panels, but there is generally no economy in reducing them.

For long spans they are built up of short pieces so placed as to break joints, or by splicing the joints, as shown in Fig. 26 *a*, or in some other suitable manner.

132.— PRATT'S TRUSS.—In the design of Pratt's Truss the chords were made of wood and of uniform size throughout, the verticals were also made of wood, and all of the same size and the diagonals were iron rods, as shown in Fig. 88.*

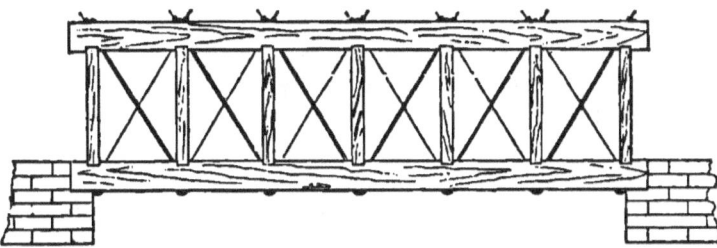

FIG. 88.

The analysis of this truss is the same as for Howe's Truss. The chords should be of the same size as for Howe's; the vertical posts may be smaller than the braces in Howe's; but the iron rods must be larger than the iron rods in Howe's. This may be the principal reason why, in wooden structures, the Howe Truss is so much more popular than Pratt's. The latter, however, is a good form, and in iron structures is much to be preferred to Howe's, as will be seen hereafter. We have used this as another " TYPE FORM."

The details of the Howe Truss—such as the use of blocks, depth of notches, etc.—are sufficiently suggestive to guide the engineer in the construction of Pratt's.

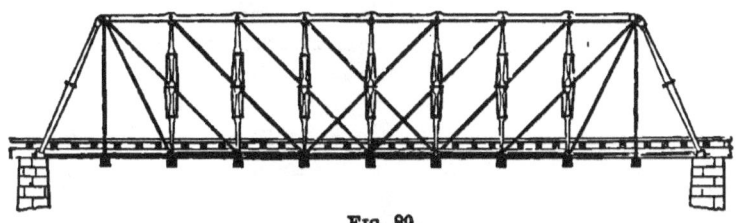

FIG. 89.

133.— WHIPPLE'S TRUSS is composed entirely of iron, in which the verticals are posts (or struts) and the diagonals are ties, as shown in Fig. 89, hence it is of the Pratt Type. This

---

* Vose's "Handbook of Rail-Road Construction."

is one of the earliest iron bridges of this country. Here we see for the first time the omission of counter diagonals. It is quite probable that in practice, in a bridge of the number of panels here shown, that one or more counter ties each side of the centre would be necessary, or at least advisable; but I have omitted them on purpose to make the application of the principle more striking. It has been frequently observed in the preceding pages that long pieces subjected to compression are liable to bend, and that in such cases their *strength* is not directly as their section, *but varies as some power of their length*, while the strength of ties varies directly as their section. For this reason it appears evident that for economy the verticals should be *posts*, and the diagonals *ties*, in iron structures of this kind.

Mr. Whipple's trusses were double, that is, each long diagonal crossed two panels. Mr Whipple so made the posts of his trusses that the tie-rods could pass through the middle of them, and at the same time he trussed the posts by iron rods in such a manner as to prevent flexure. The upper chord is usually made of hollow cast-iron tubes, the length of which equals the length of a bay. The lower chord he made of links of wrought iron, as shown in Fig. 90, which passed over cast-iron blocks. There was a cast-iron block at the foot of each post.

Since Mr. Whipple designed his truss there have been many

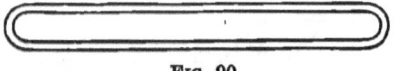

FIG. 90.

others constructed upon the same general plan, but which differ from his in the details. For instance, see Jones' Truss,[*] Murphy-Whipples',[†] Linville's Truss.[‡] Sometimes these trusses have a vertical end post instead of an inclined brace, as in Fig. 89. There is a design of such a truss by J. H. Linville, in the *Jour. Frank. Inst.*, Vol. lviii., 3d series, p. 9 (1869).

---

[*] *Scientific American*, Vol. ix., 3d series, p. 193.
[†] Col. Merrill's work on Iron Truss Rail-Road Bridges.
[‡] *Jour. Frank. Inst.*, Vol. lvii., 3d series, p. 89 (1869).

The lower chords of some of these trusses are composed of a succession of links which lap past each other, and which are connected by a pin passing through an eye near their ends. The end of one such is shown in Fig. 91.

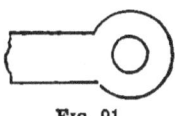

Fig. 91.

The pins which pass through the ends of the links resist shearing on both sides of the link; hence it would seem that the cross-section of the pins should equal one-half the cross-section of the links; the heads of the links being enlarged so as to equal the section of the body of the link. But experiments show that the strength of the eye depends directly upon the bearing surfaces. Experiments were made upon the links of Mr. Vignol's suspension bridge over the Dnieper, at Kieff, to determine the proper proportion for the eyes of the links and connecting pins, the results of which were given in a paper by Sir Charles Fox to the Royal Society, of which the following is an extract.*

" The links were twelve feet long from centre to centre of pinholes, and ten and a quarter inches by one inch throughout the part between the eyes. The heads were one inch thick and sixteen and one-half inches in diameter. At first these were pierced with holes four and one-half inches in diameter for the reception of the pins, thus giving the latter 15.9 square inches— or more than $1\frac{1}{2}$ times the smallest sectional area of the links, yet experiment showed that the heads would fail before the links. The heads were torn across and the metal bulged out in the direction of the strain. The head was then increased in size, but the result was substantially the same. The pin-holes in one of the original links having a head of sixteen and one-half inches in diameter were then enlarged, so as to give a semi-cylindrical bearing surface of 9.4 square inches, instead of 7 square inches as in the former case, and the strength was increased from one hundred and eighty to two hundred and forty tons, notwithstanding the diminution of the metal in the head. Subsequent experiments showed that the diameter of the pins might have been still more increased with advantage; the

---

* *Jour. Frank. Inst.*, Vol. xliii., 3d series, p. 107.

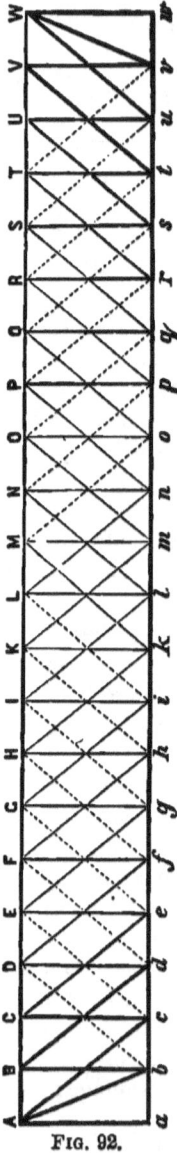

FIG. 92.

best proportions appearing to be those which give a bearing surface about equal to the least sectional area of the links. Sir Charles Fox considers that it is best to make the bearing surface slightly in excess of the proportion just mentioned, and he thus gets, for the pins of suspension chains, the simple rule "that the diameter should equal two-thirds the width of the body of the links."

This rule gives for the *hole* an excess of strength for large *pins*, and a lack of strength for small ones. In the former case the pins may be made hollow; and in the latter, strength of the pins must be computed, and leave an excess of bearing surface.

In determining the proportions of the head or eye through which the pin passes, it must be remembered that the strain is not uniformly distributed, the strain on the inner part being more severe than upon the outer, similar to the strain upon a cylinder subjected to internal pressure. (See *Resistance of Materials*, p. 26.) From the experiments made above, Sir Charles Fox estimates that the sum of the width of the two sides of the eye should be about ten per cent. greater than the width of the body of the link, both being of the same thickness throughout. But when circumstances will permit, it is better to increase the thickness of the eye. "A rule for the size of eyes which has been much used, and which gives equally good results, is to make the outside diameter of the eye equal to twice the diameter of the pin, and then to increase the thickness until the requisite sectional area is obtained.

To compute the strains on these trusses, we conceive that they are divided into two (or more if necessary) simple trusses and compute for each separately, and then combine the two results. This will be illustrated by the following example.

134.—ANALYSIS OF THE DOUBLE PANEL SYSTEM AS SHOWN IN FIG. 92.—As the end members are vertical it may be called the Linville Truss. It will require a little more material than Whipple's, which is of the trapezoidal form, because in his truss the members AB, Ab, WV, and Wv, will be omitted, and the weight of the long end braces will be but little more than the vertical posts aA and vV. In other respects the computation will be the same in the two cases.

Assume that the weight of the truss is 210,000 lbs. Also let

Span $av =$ ....200 feet $= L$
Depth of truss..........18 feet 9 in. $= D$
Length of a bay $=$ ......9$\frac{1}{21}$ feet $= l$
Number of panels...... $=$ 21 $= N$
Panel weight of one truss $=$ 5,250 $= w$
Panel weight of live load $=$ 13,640 $= p$
Inclination of the ties, $\theta = 45°$ nearly.

The weight of the truss is intended to be only that part of it which produces strains, and hence excludes the end posts and one-half of the end bays. This weight is assumed to be uniform over 20 panels, and as there are two trusses, the panel weight is 5,250 lbs.

THE CHORDS.

The maximum stress on the chords is produced by the total live and dead loads. To find the stress on PO for instance, we find it on OQ for the simple truss $Oo\ Qq\ Ss\ Uu\ Vv$; and on NP for the simple truss $Nn\ Pp\ Rr$, etc., and add the results.

A slight ambiguity arises in these cases, because if the loads be considered concentrated at the joints, and equal at each joint, they will not be symmetrically placed in reference to the centre of the truss, and hence the support at a will carry a little more than half of the load on the partial truss aA, b, D, f, etc., and the support at w will carry the same portion of the load on the other partial truss. But it simplifies the case considerably to suppose that each one sustains one-half the load for each partial truss, and I will so consider it in regard to the strain on the chords. For the sake of simplicity we will call the inclination of the diagonals 45 degrees, although it differs slightly from that. This will be equivalent to calling $2l = D$. These approximations will make the stress upon the chord somewhat less than an exact analysis, perhaps 30 or 40 lbs. less.

For the partial truss $Ww\ Uu\ Ss$, etc., let V_1 be the reaction of the support at w, and H_1 the stress on the chord.

The load will be 10 $(p + w_1)$,

$\therefore V_1 = 5\ (p + w_1)$, and taking the origin of moments at p, we have

$$H_1 D = V_1 \times 7l - 3(p + w_1) \times 4l$$

$$\therefore H_1 = \frac{23}{2}(p + w_1) = \text{stress on } PR.$$

For stress on NP, take the origin of moments at o.

$$\therefore H_2 D = V_1 \times 8l - 3(p + w_1) \times 4l$$

$$\therefore H_2 = \frac{28}{2}(p + w_1) = \text{stress on } O\ Q.$$

$$\therefore \text{Stress on } QP = H_1 + H_2 = \frac{51}{2}(p + w_1).$$

In a similar way we find the strains on all the other bays. The results are entered in the following table.

To find the weight of the hollow pieces (columns) which form the upper chords, assume that the ends are round, the column long and cylindrical, and the thickness $\frac{1}{12}$ of the external diameter. (Any other thickness might be assumed.) Then if we use *five* times the stress for safety, we may find, as in article 25, that

$$\text{Weight} = 0.00459961\ (5 \times stress)^{\frac{1}{1.83}} \times (length)^{\frac{1.79}{0.94}} *$$

$$= 0.01082713\ (stress)^{\frac{1}{1.88}} \times (length)^{\frac{1.79}{0.94}} \dots (161)$$

(log. $0.01082713 = \bar{2}.0345136$).

By means of this formula the weights in the following table have been computed.

---

* See also Col. Merrill's *Iron Truss Bridges for Rail-Roads*, p. 51.

DOUBLE PANEL SYSTEM. 157

Hence, for one-half of the truss, we have the following results:—

PIECE.	MAXIMUM COMPRESSION ON UPPER CHORD. LBS.	WEIGHT IN LBS.
$WV =$	$\frac{15}{2}(p + w_1) = 141{,}675 \ldots\ldots\ldots\ldots$	435
$VU =$	$\frac{29}{2}(p + w_1) = 217{,}235 \ldots\ldots\ldots\ldots$	547
$UT =$	$\frac{81}{2}(p + w_1) = 292{,}795 \ldots\ldots\ldots\ldots$	640
$TS =$	$\frac{37}{2}(p + w_1) = 349{,}465 \ldots\ldots\ldots\ldots$	705
$SR =$	$\frac{43}{2}(p + w_1) = 406{,}135 \ldots\ldots\ldots\ldots$	761
$RQ =$	$\frac{47}{2}(p + w_1) = 443{,}915 \ldots\ldots\ldots\ldots$	798
$QP =$	$\frac{51}{2}(p + w_1) = 481{,}695 \ldots\ldots\ldots\ldots$	834
$PO =$	$\frac{53}{2}(p + w_1) = 500{,}585 \ldots\ldots\ldots\ldots$	851
$ON =$	$\frac{55}{2}(p + w_1) = 519{,}475 \ldots\ldots\ldots\ldots$	868
$NM =$	$\frac{55}{2}(p + w_1) = 519{,}475 \ldots\ldots\ldots\ldots$	868
$ML =$	$\frac{55}{2}(p + w_1) = 519{,}475$ (half of)$\ldots\ldots$	434
Total$\ldots\ldots\ldots\ldots\ldots\ldots\ldots\ldots\ldots\ldots\ldots\ldots\ldots\ldots$		7,741

The stresses upon the lower chord are found in the same manner, observing to take the origin of moments at the proper joint of the upper chord.

The weights of the pieces are found from the formula:—

$$Weight = \frac{5 \times stress\ in\ pounds \times length\ in\ feet}{18{,}000}$$

$$= \frac{stress\ in\ pounds \times length\ in\ feet}{3{,}600}$$

This is for a factor of safety of *five*.

In this way the following table was computed:—

PIECE.	MAXIMUM TENSION ON THE LOWER CHORD. LBS.	WEIGHT IN LBS.
$v\ u = 0$		
$u\ t = \frac{5}{2}(p+w_{\scriptscriptstyle 1}) = 47,225$		125
$t\ s = \frac{15}{2}(p+w_{\scriptscriptstyle 1}) = 141,675$		380
$s\ r = \frac{23}{2}(p+w_{\scriptscriptstyle 1}) = 217,235$		575
$r\ q = \frac{31}{2}(p+w_{\scriptscriptstyle 1}) = 292,795$		775
$q\ p = \frac{37}{2}(p+w_{\scriptscriptstyle 1}) = 349,465$		920
$p\ o = \frac{43}{2}(p+w_{\scriptscriptstyle 1}) = 406,135$		1,074
$o\ n = \frac{47}{2}(p+w_{\scriptscriptstyle 1}) = 443,915$		1,174
$n\ m = \frac{51}{2}(p+w_{\scriptscriptstyle 1}) = 481,695$		1,274
$m\ l = \frac{53}{2}(p+w_{\scriptscriptstyle 1}) = 500,585$		1,324
$l\ k = \frac{55}{2}(p+w_{\scriptscriptstyle 1}) = 519,475$ (half of)		687
Total		8,308

THE TIES.

A slight ambiguity also exists in regard to the stress on the ties, but I have assumed for the dead load that it is zero at the centre, and increases uniformly to the ends; but for the live load I assume that the weights on each of the joints are equal to 13,640 lbs.

The former hypothesis makes the stress on the end ties (bA), $5\ w_{\scriptscriptstyle 1}\ sec\ \theta$, instead of $\frac{110}{21} w_{\scriptscriptstyle 1}\ sec\ \theta$.

For a maximum stress on the main ties the load must extend from the tie to the remote end, and for a counter tie, from it to the near end, as has been shown heretofore. The method of determining the stresses is sufficiently evident from previous explanations. The weights of the ties are determined from Eq (157). The results for one-half the ties are given in the following table.

DOUBLE PANEL SYSTEM. 159

PIECE.	MAXIMUM TENSION ON THE TIES. LBS.		WEIGHT IN LBS.
$b\ A =$	$\left\{\frac{110}{21}p + 5\,w_1\right\}$	$\tfrac{1}{2}\sqrt{5} = 109{,}228\ldots\ldots$	636
$c\ A =$	$\left\{\frac{100}{21}p + 5\,w_1\right\}$	$\times \sqrt{2} = 129{,}247\ldots\ldots$	952
$d\ B =$	$\left\{\frac{90}{21}p + 4\,w_1\right\}$	$\times \sqrt{2} = 112{,}365\ldots\ldots$	828
$e\ C =$	$\left\{\frac{81}{21}p + 4\,w_1\right\}$	$\times \sqrt{2} = 104{,}312\ldots\ldots$	768
$f\ D =$	$\left\{\frac{72}{21}p + 3\,w_1\right\}$	$\times \sqrt{2} = 88{,}407\ldots\ldots$	651
$g\ E =$	$\left\{\frac{64}{21}p + 3\,w_1\right\}$	$\times \sqrt{2} = 72{,}999\ldots\ldots$	538
$h\ F =$	$\left\{\frac{56}{21}p + 2\,w_1\right\}$	$\times \sqrt{2} = 66{,}285\ldots\ldots$	483
$i\ G =$	$\left\{\frac{49}{21}p + 2\,w_1\right\}$	$\times \sqrt{2} = 59{,}849\ldots\ldots$	441
$k\ H =$	$\left\{\frac{42}{21}p + w_1\right\}$	$\times \sqrt{2} = 46{,}004\ldots\ldots$	339
$l\ I =$	$\left\{\frac{36}{21}p + w_1\right\}$	$\times \sqrt{2} = 40{,}490\ldots\ldots$	299
$m\ K =$	$\left\{\frac{30}{21}p \pm 0\right\}$	$\times \sqrt{2} = 27{,}556\ldots\ldots$	203
$n\ L =$	$\left\{\frac{25}{21}p \pm 0\right\}$	$\times \sqrt{2} = 22{,}999\ldots\ldots$	169
$o\ M =$	$\left\{\frac{20}{21}p - w_1\right\}$	$\times \sqrt{2} = 10{,}917\ldots\ldots$	81
$p\ N =$	$\left\{\frac{16}{21}p - w_1\right\}$	$\times \sqrt{2} = 7{,}272\ldots\ldots$	54
$q\ O =$	$\left\{\frac{12}{21}p - 2\,w_1\right\}$	$\times \sqrt{2} =$ Negative......	
$r\ P =$	$\left\{\frac{9}{21}p - 2\,w_1\right\}$	$\times \sqrt{2} =$ Negative......	
$s\ Q =$	$\left\{\frac{6}{21}p - 3\,w_1\right\}$	$\times \sqrt{2} =$ Negative......	
$t\ R =$	$\left\{\frac{4}{21}p - 3\,w_1\right\}$	$\times \sqrt{2} =$ Negative......	
$u\ S =$	$\left\{\frac{2}{21}p - 4\,w_1\right\}$	$\times \sqrt{2} =$ Negative......	
$v\ T =$	$\left\{\frac{1}{21}p - 4\,w_1\right\}$	$\times \sqrt{2} =$ Negative......	
	Total...............................		6,442

The Vertical Struts.

The vertical component of the stress on the struts is the same as for the ties, as has been before shown, and hence the general expressions are given in the parenthesis of the preceding table. Their weights are computed from Eq. (156). The results for half the struts are given in the following table:—

PIECE.	MAXIMUM COMPRESSION ON THE POSTS. LBS.	WEIGHT IN LBS.
$a\ A =$	$10\,p + 10\,w_1 = 188{,}900\ldots$	1,841
$b\ B =$	$\frac{90}{21}p + 4\,w_1 = 79{,}457\ldots$	1,162
$c\ C =$	$\frac{81}{21}p + 4\,w_1 = 73{,}621\ldots$	1,115
$d\ D =$	$\frac{72}{21}p + 3\,w_1 = 62{,}514\ldots$	1,024
$e\ E =$	$\frac{64}{21}p + 3\,w_1 = 57{,}319\ldots$	977
$f\ F =$	$\frac{56}{21}p + 2\,w_1 = 49{,}393\ldots$	902
$g\ G =$	$\frac{49}{21}p + 2\,w_1 = 44{,}531\ldots$	854
$h\ H =$	$\frac{42}{21}p + w_1 = 34{,}420\ldots$	745
$i\ I =$	$\frac{36}{21}p + w_1 = 30{,}253\ldots$	695
$k\ K =$	$\frac{30}{21}p + 0 = 19{,}486\ldots$	550
$l\ L =$	$\frac{25}{21}p + 0 = 16{,}238\ldots$	499
	Total = $\ldots$	10,364

Hence we have:—   lbs.
Weight of upper chord = $\ldots$ 7,741
Weight of lower chord = $\ldots$ 8,308
Weight of ties = $\ldots$ 6,442
Weight of posts = $\ldots$ 10,364

Total = $\ldots$ 32,855
               2

Multiplied by 2, weight of one truss = $\ldots$ 65,710

The computation thus far is for a mere *skeleton truss*, and hence something must be added for connections, pins, bolts, cross-ties, floor and track, to find the full weight of the *bridge*. Some of these might be computed, but instead of attempting it, we will add 15 per cent. of the above weights for the mechanical connection, and add *

For iron floor beams$\ldots$ 100.80 lbs. per foot.
Top lateral struts$\ldots$ 21.60 lbs. per foot.
Top lateral ties$\ldots$ 24.00 lbs. per foot.
Track (including rails, ties, etc., etc.)$\ldots$ 245.60 lbs. per foot.

Total$\ldots$ 392.00 lbs. per foot.

* I have taken some of these numbers so as to make them agree with those in an article which I published in the *Rail-Road Gazette*, Dec. 24th, 1870.

And for half the bridge we multiply this by 100.
Hence we have :—
Weight of one truss.......................... 65,710 lbs.
Weight 15 per cent........................... 9,857 lbs.
Half weight of floor, etc., etc., as above......... 39,200 lbs.

Weight of half the bridge...............114,767 lbs.

Total weight of the bridge.................229,534 lbs.

After deducting the weight of the end posts, half the weight of the end ties, and half the weight of the end bays of the chords, I find that the computed weight exceeds the assumed weight by about 5,000 lbs. This, however, would increase the preceding result by a small amount only.

I have also made the following computations, using the same live load per foot of length as above, and assuming that the ties incline at an angle of 45°.

Live Load per Panel.	No. of Panels.	Assumed weight of the Truss.	Depth.	Weight of Upper Chord.	Weight of Lower Chord.	Weight of the Ties.	Weight of the Posts.	Computed Weight of the Truss.
Lbs.		Lbs.	Ft.	Lbs.	Lbs.	Lbs.	Lbs.	Lbs.
17,600	16	240,000	25.00	8,895	6,221	6,790	13,748	242,408
14,080	20	240,000	20.00	7,953	7,934	6,615	11,133	233,234
13,640	21	280,000	18.75	7,884	8,935	10,449	10,748	232,753
13,640	21	210,000	18.75	7,741	8,308	10,364	6,442	229,534

The live load is nearly the same per foot of length in all these cases. We see that the assumed weight has much less influence upon the resultant weight than the depth of the truss has. The data in the first two cases are essentially the same except the depth, and we see that the resultant weight is less for the less depth. This result appears still more striking by comparing the third case with the others; for here the weight is 40,000 lbs. more, and the depth is 15 inches less, than in the preceding case, and yet the computed weight is less than in the preceding case.

If the members which are subjected to compression are proportioned as for pillars, it is easy to show that for minimum material the depth must be infinitely small.

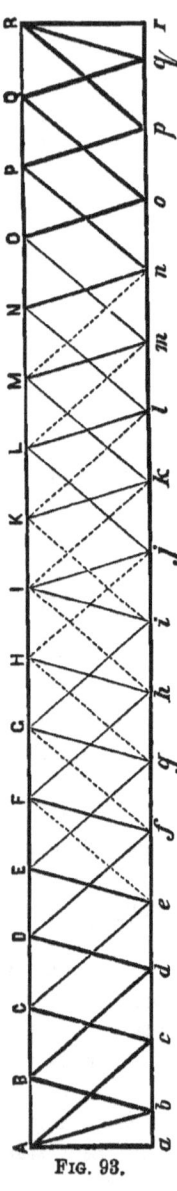

FIG. 93.

One of the builders of these trusses (and perhaps many others do the same) insists that a depth equal to one-eighth of the span gives the best results. As we have just seen, it does not give the *minimum material*, but it must be observed that the less the depth the greater is the number of bays, and hence more pieces must be used. The small saving in material in a low truss is more than balanced by the extra cost of labor in manufacturing them. Indeed the whole problem of "Minimum Material" in trusses is more theoretical than practical. The saving of a few lbs. of material is as nothing compared with the best modes of connecting the parts, and of conveniences in erecting the structure. It would also be folly—if not madness—to insist upon saving a few pounds for the sake of making a lighter truss than some other builder, if by so doing it involved any known risk of safety or durability.

135.—POST'S TRUSS is a slight modification of the ordinary panel system. The posts incline towards the centre having a run of half a bay; the ties cross a post and incline 45°, and the counter-ties have the same inclination, but cross only one panel, as shown in Fig. 93.

The mode of analysis is essentially the same as that already explained for the panel system. Let

Span ar = ... 200 feet = L
Depth of truss .... 18 feet 9 in. = D
No. of panels .... 16 = N
Panel length .... 12 feet 6 in. = l
Panel weight of engine. 17,600 lbs. = p
Panel weight of one truss 6,562½ lbs. = w

This load is essentially the same as that

assumed in the preceding problem. The span and depth are the same.

TIES.

To find the maximum strain on the ties and posts, suppose that the live load extends over the whole length of the truss, and moves off, without shock, in the direction from r towards a. The maximum strain on Rr and Rq will be when the load extends the whole length; on Rp when the load extends from p to a; on oQ and Qq when it extends from o to a, and so on to the centre, beyond which we get the maximum on the counter-ties only; and to get the maximum on the other half of the bridge, we suppose that the live load moves in the opposite direction, and consider the maximum on the post and main ties to the centre, and on the counter-ties beyond the centre, as before. The results are entered in the annexed table. To show how they are obtained, take cD, for example, and consider first the *moving load*. Since ab is only half a bay, the joint b will sustain only $\frac{1}{2}p$ (although in considering the dead load I have supposed that w_1 is supported there). If the truss had no weight the loads at b and c would produce tension on a diagonal passing from c to D. Although this diagonal is not shown, yet we may find the stress which tends to pass through it.

The strain is produced by a load of $\frac{1}{2}p$ at b, and p at c. A portion of the strain due to $\frac{1}{2}p$ is transmitted through bC, thence to c, thence through cD (not shown in the figure), and so on to r. Similarly the strain due to p at c is transmitted through cD, and so on to r. Hence, if we find how much of the weight is sustained at r, we find how much would be transmitted through cD, by multiplying the result by the secant of the inclination. This is easily found by moments. Taking the origin of moments at a, the lever arm of $\frac{1}{2}p$ is $ab =$ one-half a bay; of p at c, is $ac = 1\frac{1}{2}$ bays; and the arm of the moment of the reaction at r is 16 bays.

Hence we have $16 \times \text{reaction} = \frac{1}{2}p \times \frac{1}{2} + p \times \frac{3}{2}$

$\therefore \text{reaction} = \frac{15}{128}p$

In a similar way find the stress due to the weight of the truss, by observing that it is loaded the whole length by the weight of the truss. Or, observe that the strain at the middle due to the weight of the truss is zero, and increases each way from the centre, and we find that the stress due to w_1 is $3w_1 \sec \theta$. Hence the resultant stress due to both loads is

$$\left(\tfrac{15}{128} p - 3w_1\right) \sec \theta,$$

as shown in the following table. In this case it is negative. The secant of the inclination is $\sqrt{2}$, and the length of a tie is $18.75 \times \sqrt{2}$. Hence we readily find the following results:

Piece.	Maximum Tension on the Ties. LBS.	Weight in LBS.
$q\ R =$	$\left[\tfrac{1053}{256} p + 4 w_1\right] \tfrac{1}{8} \sqrt{10} = 103,979$	571
$p\ R =$	$\left[\tfrac{991}{256} p + 4 w_1\right] \times \sqrt{2} = 127,542$	866
$o\ Q =$	$\left[\tfrac{867}{256} p + 3 w_1\right] \times \sqrt{2} = 112,138$	826
$n\ P =$	$\left[\tfrac{699}{256} p + 3 w_1\right] \times \sqrt{2} = 95,804$	706
$m\ O =$	$\left[\tfrac{651}{256} p + 2 w_1\right] \times \sqrt{2} = 81,856$	603
$l\ N =$	$\left[\tfrac{499}{256} p + 2 w_1\right] \times \sqrt{2} = 67,078$	494
$k\ M =$	$\left[\tfrac{407}{256} p +\ w_1\right] \times \sqrt{2} = 54,685$	403
$j\ L =$	$\left[\tfrac{331}{256} p +\ w_1\right] \times \sqrt{2} = 37,927$	280
$i\ K =$	$\left[\tfrac{315}{256} p +\ 0\right] \times \sqrt{2} = 30,627$	226
$h\ I =$	$\left[\tfrac{195}{128} p -\ w\right] \times \sqrt{2} = 28,625$	211
$g\ H =$	$\left[\tfrac{148}{128} p - 2 w_1\right] \times \sqrt{2} = 9,283$	69
$f\ G =$	$\left[\tfrac{99}{128} p - 3 w_1\right] \times \sqrt{2} =$ Negative....	
$e\ F =$	$\left[\tfrac{63}{128} p - 4 w_1\right] \times \sqrt{2} =$ Negative....	
$d\ E =$	$\left[\tfrac{35}{128} p - 5 w_1\right] \times \sqrt{2} =$ Negative....	
$c\ D =$	$\left[\tfrac{15}{128} p - 6 w_1\right] \times \sqrt{2} =$ Negative....	
$b\ C =$	$\left[\tfrac{3}{128} p - 7 w_1\right] \times \sqrt{2} =$ Negative....	
	Total...................................	5,467

POSTS.

The end posts sustain the weights on both partial trusses, which equals the sum of the quantities in the first two parentheses of the preceding table, multiplied by the secant of the inclination, which is $\tfrac{1}{8}\sqrt{10}$. The length of a post is $\tfrac{1}{8}$ of $18.75 \sqrt{10}$.

When the train extends from the centre to one end of the

bridge, it is impossible to tell exactly how much is sustained by Ii Ij respectively, but the most rational assumption is, that each transmits half the strains from I to the lower chord. We readily find the following results:—

Piece.	Maximum Compression on the Posts. LBS.				Weight in LBS.
$r\ R =$	$\tfrac{1964}{256}p + 8\,w_1$			$= 188{,}900\ldots$	1,842
$q\ Q =$	$[\tfrac{897}{256}p + 3\,w_1]$	$\tfrac{1}{8}$	$\sqrt{10}$	$= 81{,}335\ldots$	1,498
$p\ P =$	$[\tfrac{609}{256}p + 3\,w_1]$	$\tfrac{1}{8}$	$\sqrt{10}$	$= 69{,}160\ldots$	1,198
$o\ O =$	$[\tfrac{651}{256}p + 2\,w_1]$	$\tfrac{1}{8}$	$\sqrt{10}$	$= 59{,}513\ldots$	1,102
$n\ N =$	$[\tfrac{499}{256}p + 2\,w_1]$	$\tfrac{1}{8}$	$\sqrt{10}$	$= 48{,}498\ldots$	988
$m\ M =$	$[\tfrac{467}{256}p + \ w_1]$	$\tfrac{1}{8}$	$\sqrt{10}$	$= 40{,}011\ldots$	893
$l\ L =$	$[\tfrac{331}{256}p + \ w_1]$	$\tfrac{1}{8}$	$\sqrt{10}$	$= 29{,}195\ldots$	755
$k\ K =$	$[\tfrac{315}{256}p + \ 0]$	$\tfrac{1}{8}$	$\sqrt{10}$	$= 22{,}827\ldots$	662
$j\ J =$	$[\tfrac{195}{256}p + \ 0]$	$\tfrac{1}{8}$	$\sqrt{10}$	$= 14{,}131\ldots$	512
	Total$\ldots\ldots$				9,450

THE CHORDS.

The maximum strain on the chords is produced by a load extending the whole length of the span. The strain on any bay is most easily found by the principle of moments. For example, take no. If this piece be severed the truss may fall by turning about O or P. To find the strain on op, suppose that the truss is separated into two simple ones, one being composed of the parts rR, Rp, pP, Pn, etc.; and the other of rR, Rq, qQ, Qo, oO, Om, etc. We find the strain on np for the first partial truss, and oq for the second, and add the results for the strain on op. Let V_1 be the amount sustained at r for the first truss, and V_2 that sustained by the second truss. By examining the table of strains for the ties, we see that the reaction at a for the first partial truss is $\tfrac{1053}{256} p + 4\,w_1$. The origin of moments being at Q, let fall a perpendicular Qx, then will the lever arm of V_1 be $l = xr$.

The load at q being $\tfrac{3}{4}p + w_1$, and its arm being $x\,q = \tfrac{1}{2}\,l$, its moment will be $(\tfrac{3}{4}\,p + w_1)\,\tfrac{1}{2}\,l$; and as V_1 and this load act in opposite senses, we have for the total moment,

$$H_1 D = V_1 l - (\tfrac{3}{4} p + w_1) \tfrac{1}{2} l; \text{ but } D = \frac{3}{2} l$$

$$\therefore H_1 = \left[\left(\frac{1053}{256} p + 4 w_1 \right) - \tfrac{1}{2}(\tfrac{3}{4} p + w_1) \right] \tfrac{2}{3} =$$

$$\frac{957}{384} p + \frac{7}{3} w_1$$

For the other partial truss, let fall a perpendicular from P to the lower chord. Calling the strain on $n\, p$ for this part of the truss H_2, and we find,

$$H_2 D = \left[\left(\frac{931}{256} p + 4 w_1 \right) 2 l - (p + w_1) \tfrac{1}{2} l \right]$$

$$\therefore H_2 = \frac{1734}{384} p + \frac{15}{3} w_1$$

Hence the strain on $o\, p$ is

$$H_1 + H_2 = \frac{2691}{384} p + \frac{22}{3} w$$

as given in the table.

In a similar way the strains on any part of the upper or lower chord are found and entered in the table. The length of each regular bay, which is the length of the pieces of the chord, is $12\tfrac{1}{2}$ feet.

Piece.	Maximum Tension on the Lower Chord. LBS.	Weights. LBS.
$a\, b$ or $q\, r$	0	
$b\, c$ or $q\, p =$	$\frac{957}{384} p + \frac{7}{3} w_1 = 59{,}174$	205
$c\, d$ or $p\, o =$	$\frac{2691}{384} p + \frac{22}{3} w_1 = 171{,}462$	595
$d\, e$ or $o\, n =$	$\frac{4584}{384} p + \frac{33}{3} w_1 = 282{,}287$	980
$e\, f$ or $n\, m =$	$\frac{5763}{384} p + \frac{44}{3} w_1 = 360{,}387$	1,257
$f\, g$ or $m\, l =$	$\frac{6589}{384} p + \frac{51}{3} w_1 = 413{,}558$	1,435
$g\, h$ or $l\, k =$	$\frac{7599}{384} p + \frac{58}{3} w_1 = 475{,}162$	1,650
$h\, i$ or $k\, j =$	$\frac{7677}{384} p + \frac{61}{3} w_1 = 485{,}200$	1,085
$j\, i =$	$\frac{8965}{384} p + \frac{64}{3} w_1 = 509{,}646$ (half)	885
Total		8,686

POST'S TRUSS.

PIECE.	MAXIMUM COMPRESSION ON THE UPPER CHORD. LBS.	WEIGHTS LBS.
$RQ =$	$\frac{1282}{256}p + \frac{16}{8}w_1 = 119{,}805$	668
$QP =$	$\frac{2430}{256}p + \frac{28}{8}w_1 = 228{,}414$	943
$PO =$	$\frac{3330}{256}p + \frac{40}{8}w_1 = 326{,}437$	1,139
$ON =$	$\frac{4137}{256}p + \frac{48}{8}w_1 = 389{,}420$	1,249
$NM =$	$\frac{4696}{256}p + \frac{56}{8}w_1 = 445{,}350$	1,347
$ML =$	$\frac{5161}{256}p + \frac{60}{8}w_1 = 487{,}068$	1,409
$LK =$	$\frac{5373}{256}p + \frac{64}{8}w_1 = 499{,}738$	1,428
$KT =$	$\frac{5502}{256}p + \frac{64}{8}w_1 = 508{,}273$	1,441
	Total =	9,624

The sum of these gives the weight of one-half the truss. We have

Weight of ties = ... 5,467 lbs.
Weight of posts = ... 9,450 lbs.
Weight of lower chord = ... 8,686 lbs.
Weight of upper chord = ... 9,624 lbs.

Total .... 33,227 lbs.

Multiply by two and we have weight of one truss = ... $\frac{2}{66{,}454 \text{ lbs.}}$

Allowing 15 per cent. on this weight for connections, as before, and the same quantities as in the preceding case for beams, etc., as follows, and we have,

For iron floor-beams............... 100.80 lbs. per foot.
Top lateral struts................... 21.60 lbs. per foot.
Top lateral ties..................... 24.00 lbs. per foot.
Track (including rails, ties, etc.)...... 245.60 lbs. per foot.

Total........................ 392.00 lbs. per foot.

And for half the bridge we multiply this by 100. Hence we have for

POST'S TRUSS.

Weight of one truss .... 66,454 lbs.
Weight 15 per cent................. 9,968 lbs.
Weight of floor, etc., as above........ 39,200 lbs.

Weight of half the bridge..... 115,622 lbs.

Total weight of bridge .... 231,244 lbs.

This result is about 1,700 lbs. greater than for the panel system of the same data. This result in kind might have been anticipated, since posts being much heavier than the ties, should not incline so much as this system demands (18° 26′ from the vertical, since *cotang.* of inclination $= \frac{3}{1} l \div \frac{1}{3} l = 3$), as was shown in article 122. Still, in the eyes of some, the Post truss may have other advantages which more than overbalance this loss (if we may so call it) of metal. If they have the same depth and the ties make two intersections—or in other words, cross two panels—and the ties incline at an angle of 45 degrees, as they do in practice in both systems, then the number of bays and panels in the panel system is about 25 per cent. greater than the Post truss. In the example which we have analyzed, the Post truss has 16 panels and the other truss 21 panels, so that for the same span and depth the Post truss has fewer posts, fewer ties, and a less number of parts in the chords; the length of the chords is the same, the length of the ties the same, and the length of the posts $\frac{1}{3} \sqrt{10} = 1.054$ longer than the corresponding parts of the other truss. A comparison of the results shows that the total weight of the ties and posts in the panel truss exceeds considerably the total weight of the same pieces in the Post truss. This excess, however, is more than made up by the greatly diminished weight of the upper chord, which is caused by the shorter pieces of which it is composed, in the example before us. As we have seen, the weight increases nearly as the square of the length of the compression members.

We see that the difference in weight is quite too small to establish the superiority of one over the other. The simplicity of the details, the ease with which they may be erected, their

POST'S TRUSS.   169

liability to get out of repair, and the ease with which they may be repaired, are more vital questions than the mere saving of a few lbs. of iron. Still the question of form is a very important one. Lightness in itself is desirable, but other qualities are equally desirable, and the engineer should seek to combine as many good qualities as possible.*

---

* Mr. Post gave this inclination to the posts so as to secure less material than in the ordinary panel system with vertical posts. An attempt to prove this is given in Col. Merrill's "Iron Truss Bridges for Rail-Roads," p. 121, but there are several assumptions in the demonstration which are not demanded by the problem, and, by introducing them, make the solution of little or no value. In the first place it is *assumed* that the middle pair of posts meet in the middle, as at I, Fig. 92. This is not necessary in the discussion, although the inventor may construct it in that way.

In the next place it is *assumed* that the run of a brace is half a bay. This is begging the question; for he does not consider the run of a tie in the solution, and hence it would be just as fair to assume that the run is one-fourth, or any other fraction of a bay. In other words, the number of bays is independent of the run of a brace. Removing these two *assumptions* from the solution, and it follows quickly that the posts should be vertical. For we have

$W=$ the weight to be carried;
$h=$ the depth of the truss; and
$b=$ the run of the brace.

$$\therefore \text{Stress on the brace} = \frac{W\sqrt{b^2+h^2}}{2h}$$

If $N=$ the number of braces, their

$$Volume = constant \times N\left[\frac{W\sqrt{b^2+h^2}}{2h}\right]^m \times \left[\sqrt{b^2+h^2}\right]^n$$

in which $m = \frac{1}{1.88}$, and

$$n = \frac{1.79}{0.94}$$

As b is here the only variable, it is evidently a minimum for $b=0$.

But if N is a function of the run of the brace, as, for instance, $N = \frac{L}{cb}$, where c is an assumed constant, and $L=$ the span, we have,

$$Volume = constant \times \frac{L}{cb}\left[\frac{W\sqrt{b^2+h^2}}{2h}\right]^m \times \left[\sqrt{b^2+h^2}\right]^n$$

$$= constant \times \frac{L}{c}\left[\frac{W}{2h}\right]^m \times \frac{(b^2+h^2)^{\frac{m+n}{2}}}{b}$$

which is to be a minimum. All is constant but the last factor. Solving gives $b = 0.8336\, h$

$$\therefore \frac{b}{h} = 0.8336 = tang\ 39°\ 49'$$

In this solution the number of braces is arbitrary, and their run may be any fractional part of the bay.

136.—EXTRA STRAINS.—We have supposed that the chords were parallel and horizontal. But if they sag, or are arched, the strains upon the several parts will be slightly modified. For instance, if they sag, the strains in the chords will have a downward component. This component may produce an additional stress on the verticals. But I think that in no case of good workmanship will it be necessary to consider the effect of such a stress.

MULTIPLE SYSTEMS.

137.—HAUPT'S LATTICE, is of the form of a multiple panel system, as shown in Fig. 93. It is a wooden structure in

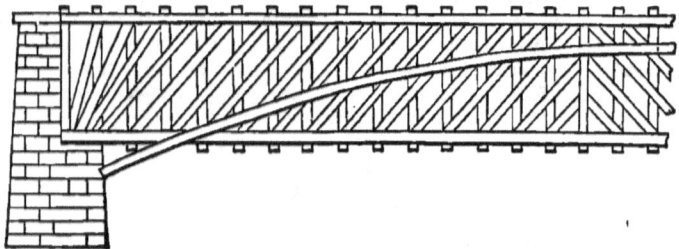

FIG. 93.

which one-half the pieces are vertical and serve as ties, and the other half are inclined at an angle of about 45 degrees and serve as braces. There are no counter-braces. The main braces meet in the centre, as shown in the figure.

For heavy loads on a light structure this would be a weak point. By securing the main braces at their ends to the verticals, or to the chords, would cause them to act as counter-braces when necessary. The inventor supposed that one-half the pieces should be vertical in order to resist the vertical forces in the truss, but we have seen in the triangular system that this is not necessary, and it is questionable whether it possesses as much advantage over Towne's Lattice as was at first supposed.

The inventor introduced an arch into the system, which makes it a compound system. The arch will add very much to the stiffness of the structure, as well as to its strength. Without the arch, it may be analyzed in the same way as Towne's Lattice, but with the arch it is impossible to make an exact analysis.

138.—HALL'S LATTICE.—If Haupt's Lattice be inverted we shall have the mechanical conditions of Hall's Lattice. The

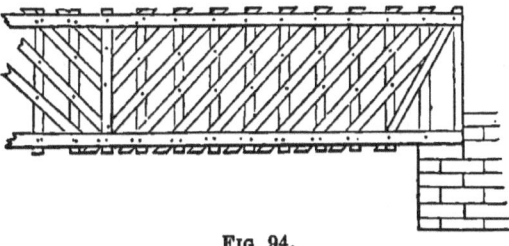

FIG. 94.

inventor was induced to make it of this form because the ultimate resistance of wood is greater for tension than for compression, and the oblique pieces are subjected to greater strains than the verticals. But as the pieces cannot hold more than their fastenings, and as it is difficult to secure the ends of wooden pieces so as to bring into use their full tensile strength, it is doubtful if this is any improvement over other forms of lattice trusses.

139.—LATERAL HORIZONTAL BRACING is necessary to resist the side pressure of the wind and to prevent swaying from passing loads. The pressure of the wind is like a uniform load, and its amount may be determined with sufficient accuracy by multiplying the total side area in square feet which is exposed to the wind by 40; the result will be the pressure in lbs.

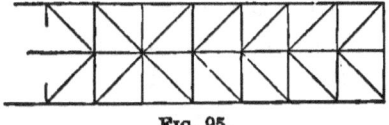

FIG. 95.

The bracing must resist the pressure from both sides, and may be like a Howe Truss placed under the roadway or over it, or both under and over, or it may be arranged as indicated in Fig. 95.

140.—KNEE BRACES are to hold the side trusses elect against the pressure of the wind. The trusses are placed on each side of the roadway and the latter is suspended from them. When necessary a knee-brace A, Fig. 96, or a tie-brace B, of the same figure, is introduced. The total pressure of the wind being known, and its centre of pressure being at half the depth of the truss, and the truss tending to turn about b, Fig. 96, it is

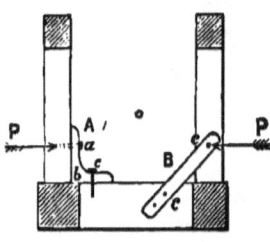
Fig. 96.

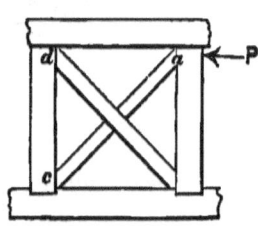
Fig. 97.

easy to find the pressure P upon the knee brace at the point where the bolt a is inserted. Similarly for the point e on the other side. The dimensions of the knee brace may then be found on the condition of a beam fixed at one end, and a force P applied at the other. The stress on B will be $P \sec \theta$.

In the case of a *deck bridge*, or one in which the whole bridge is below the roadway, there will be room to place the braces as in Fig. 97. In such cases the braces need not be large.

141.—STABILITY OF THE BRIDGE ON ITS SUPPORTS. —The bridge may be overturned by the force of the wind, or it may be slid off by the same force. If

$A =$ the area in feet exposed to the wind;
$B =$ the width of the bridge on the supports;
$W =$ the total weight of the bridge; and
$f =$ the force of the wind per square foot.

we must have, in order to prevent overturning,

$$W \times \tfrac{1}{2} B > f A \times \tfrac{1}{2} D.$$

f in ordinary cases rarely exceeds 8 or 10 lbs. per square foot, but in rare cases it has exceeded 40 lbs. This will be

LOADED BEAM. 173

more fully discussed in the Part on Roofs. If the above inequality does not exist, the bridge should be anchored in some suitable way.

If $f =$ the coefficient of friction, we must have, to prevent slipping,

$$fW > f A$$

Values of f

For Oak upon oak............$f = 0.62 = \frac{5}{8}$ nearly
  Oak upon elm............$f = 0.38 = \frac{1}{3}$ nearly
  Iron upon oak............$f = 0.65 = \frac{2}{3}$ nearly
  Cast iron upon cast iron....$f = 0.16 = \frac{1}{6}$ nearly
  Oak upon calcareous stone.$f = 0.49 = \frac{1}{2}$ nearly

If the above inequality does not exist, the bridge must be anchored.

GRAPHICAL REPRESENTATION OF THE LAW OF STRAINS.

142.—CONTINUOUS LOADING—VERTICAL SHEARING STRESS.—Let a beam AC, Fig. 98, be loaded uniformly, and continuously from one end A to some point B. Let the beam be considered as a uniform load throughout. If we conceive that a truss has indefinitely short bays, so that if one of the chords be severed it may turn about

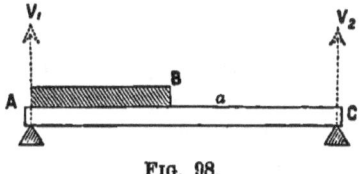

FIG. 98.

any *point*, considered as a joint, the case will be essentially the same as the one we are considering; and even when the bays have a finite length, the formulas which we shall develop from this case will be sufficiently exact for most cases.

The chief advantage of assuming that the strains are continuous, instead of being concentrated at joints (or *nodes*) only, as in the preceding cases, is that the *law* of change can be represented by straight lines, or by continuous curves, as the case may be, and the results may be more easily discussed.

Let $L = AC =$ distance between the supports;
  $x = AB =$ the length of live load;

$w =$ the weight per foot of length of the beam (dead load);
$w' =$ the weight per foot of length of the live load;
$W_1 = wL =$ the weight of the beam or bridge;
$W_2 = w'L =$ the total weight of the live load when it extends the whole length of the span;
$W = W_1 + W_2$;
$V_1 =$ the reaction of the support at A;
$V_2 =$ the reaction of the support at C;
$S_s =$ the shearing stress; and
$z = Ca =$ any distance measured from C.

We readily find that

$$V_1 = \tfrac{1}{2} wL + \frac{w'}{L}(L - \tfrac{1}{2}x)\, x \ldots\ldots\ldots\ldots (163)$$

$$V_2 = \tfrac{1}{2} wL + \frac{w'x^2}{2L} \ldots\ldots\ldots\ldots\ldots\ldots (164)$$

The vertical shearing stress at any point between C and B is found by subtracting from the value of V_2 all the load between the end C and the joint considered.

$$\therefore S_s = V_2 - wz = \tfrac{1}{2} wL + \frac{w'x^2}{2L} - wz \ldots (165)$$

Now we desire to show that the shearing at the point a is greatest when the load extends from A to a (or $x + z = L$) and Aa is greater than aC, or $x > z$.

1st, Suppose that the load is uniform on Ba, and equal to $w' \times Ba$, then we readily find that the shearing stress at a is

$$\frac{w' \times Ba\,(x + \tfrac{1}{2} Ba)}{L} + \tfrac{1}{2} wL + \frac{w'x^2}{2L} - wz,$$

which is greater than Eq. (165.) It is also evident that if any part of Ba be loaded, the shearing stress at a will be greater than if it be wholly unloaded.

2d, Suppose that the load extends from A to some point to the right of a. Then a fractional part of this load, say $\frac{1}{n}$-th part, will be supported at C, but in obtaining the shearing stress, the *whole* of the load at the right of a must be deducted from the reaction of the support at C; and hence the shearing stress at a will be less than if it extended only to a.

If the load extends from A to a, $z = L - x$, and Eq. (165) becomes

$$S_s = \tfrac{1}{2} wL + \frac{w' x^2}{2 L} - w(L - x)\ldots\ldots\ldots(166)$$

3d, Suppose that the live load extends only from a to C, then will the shearing stress at a be

$$S'_{s'} = V_1 - wx = \tfrac{1}{2} wL + \frac{w'}{2 L}(L - x)^2 - wx,$$

which taken from Eq. (166) gives

$$S_s - S^{*}{}_{s'} = -\tfrac{1}{2}(w' + 2w)(L - 2x);$$

which is zero for $x = \tfrac{1}{2} L$;
Negative for $x < \tfrac{1}{2} L$;
Positive for $x > \tfrac{1}{2} L$; hence
the vertical shearing stress at any point for a uniform load is greatest when the greater segment of the span is loaded, and the shorter unloaded.

For this case Eq. (166) becomes

$$S_s = \frac{w' x^2}{2 L} + w x - \tfrac{1}{2} wL = y \text{ (say)} \ldots\ldots\ldots(167)$$

Equation (167) is the Equation of a parabola, in which y is vertical, as shown in Fig. 98. The ordinates of the curve BD represent the vertical shearing stress at the rear end of the live load for uniform live and dead loads, the length of the live load being equal to the length of the span, and moving off from the beam towards the left, or moving on to the beam towards the right, without shock. The curve through E represents a similar vertical shearing stress for a live load moving in the opposite direction.

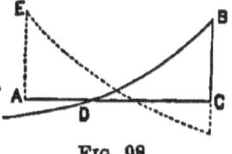

Fig. 98.

In truss bridges with parallel chords the diagonals incline one way from C to D, and in the opposite direction from A to where the dotted line crosses AC, and the middle part of the line which is included between the two curves represents the part of the truss in which diagonals incline both ways.

For $x = L$, y is a maximum $= BC =$
$$\tfrac{1}{2}(w + w') L = \tfrac{1}{2} W.$$

To find where the curve crosses the axis of x, make $y = 0$ in Eq. (167) and solve for x.

$$\therefore x = x_o = \left[-1 \pm \sqrt{1 + \frac{w}{w'}} \right] \frac{w}{w'} L \ldots \ldots (168)$$

This value (x_o) is also the point of zero shearing stress.
If $w = w'$; $x_o = 0.41\ L$ or $-2.41\ L$
$2\ w = w'$; $x_o = 0.36\ L$ or $-1.18\ L$
$5\ w = w'$; $x_o = 0.29\ L$ or $-0.46\ L$
$10\ w = w'$; $x_o = 0.23\ L$ or -0.45.

The negative values do not come within the limits of the problem and only have an analytical signification.

The value of x_o shows how far from each end counter-braces are unnecessary, and as the live load rarely exceeds two or three times the dead load, it appears that we may safely say in practice that for a distance of $0.3\ L$ from each end no counter-braces are needed.

The effect of the live and dead loads may be shown separately. Thus, in Eq. (167) let

$$\frac{w'x^2}{2L} = y_1$$

and

$$wx - \tfrac{1}{2}wL = -y_2.$$

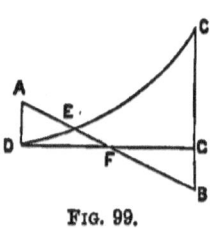

Fig. 99.

the former of which is for the live load, and is the Equation of a parabola, as in Fig. 99, in which the vertex of the curve is at the origin of co-ordinates, at D. The latter is for the dead load, and is represented by the straight line AB. The point E where these lines cross is the point of zero shearing.

143.—GENERAL PROBLEMS.—1. Find a general expression for the point of vertical zero shearing stress, when a beam has a uniform dead load and a uniform live load, the live load extending from one end to any point of the span.

2. Find a general relation between the centre of gravity and the point of zero shearing stress for a uniform dead load and a uniform live load, the latter of which extends from one end to any point of the span.

LAW OF STRAINS UPON THE CHORDS.

3. Show from the preceding when the distance between the point of zero shearing and the centre of gravity is a maximum, and when it is a minimum.

4. Suppose that a uniform load, as a train of cars, is headed by a single weight P, as a locomotive, and that there is a uniform dead load: required the expression for the shearing stress at the point where P is applied as the train moves along, supposing that all of P is applied at a point.

144.—STRAINS UPON THE CHORDS.—If we suppose that the chords are liable to break at any point, instead of at a joint,

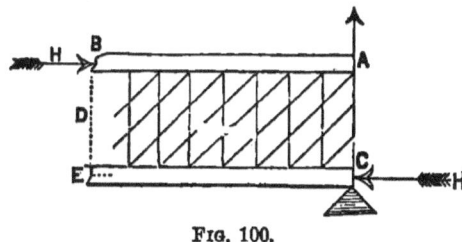

FIG. 100.

we find a very simple expression for the strains upon the chords.

In Fig. 100,
Let $x = AB$;
$D = BE =$ the depth;
$V = \frac{1}{2} W =$ one-half the total load;
$H =$ the strain upon the chord at any point;
$w =$ the load per foot of length of live and dead loads.
Taking the origin of moments at E, and we have

$$\tfrac{1}{2} Wx - \tfrac{1}{2} wx^2 = H.D \quad \ldots \ldots \ldots \ldots (169)$$

$$\therefore H = \frac{Wx - wx^2}{2D} \quad \ldots \ldots \ldots \ldots (170)$$

which is the equation of a parabola, whose axis is vertical, and whose origin is over the middle of the span, as shown in Fig. 101. The value of H is a maximum for $x = \tfrac{1}{2} L$; for which $H = \tfrac{1}{8} \dfrac{WL}{D}$.

FIG. 101.

It also diminishes as the depth increases.

PROBLEMS.—1. What must be the depth of the truss so that the stress in the chords at the middle of the span shall equal $\frac{1}{n}th$ part of the total live and dead loads?

2. Find the point where the strain upon the chords equals the vertical shearing stress.

NOTE.—The vertical shearing stress multiplied by the *sec. θ* gives the stress upon the diagonals.

3. Find the point where the stress upon the diagonals equals the stress upon the chords.

145.—RELATION BETWEEN THE SHEARING STRESS AND THE MOMENTS OF APPLIED FORCES.—By differentiating Equation (169), we have

$$\frac{d(HD)}{dx} = \tfrac{1}{2} W - w x \ldots \ldots \ldots \ldots (171)$$

the second member of which is the value of the shearing stress at any point for a uniform permanent load. Hence, *the shearing stress is the first differential coefficient of the moment of applied forces* (the forces being perpendicular to the axis of the beam).

The reverse is also true, that the moments of applied forces may be found by multiplying the expression for the shearing stress by the differential of the abscissa and integrating the expression.

CHAPTER IV.

TRUSSES WITH CHORDS NOT PARALLEL.

146.—McCALLUM'S TRUSS.—One of the simplest bridges of this kind is called McCallum's Truss, as shown in Fig. 102. This is a view of one-half of the Susquehannah Bridge on the New York and Erie Railroad. The span is nearly 200 feet.* The *general style* of this truss is that of the "Howe Type." The lower chord is horizontal. The *main peculiarity* is the curvature of the upper chord, although the inventor gave due consideration to the details of the construction. The long

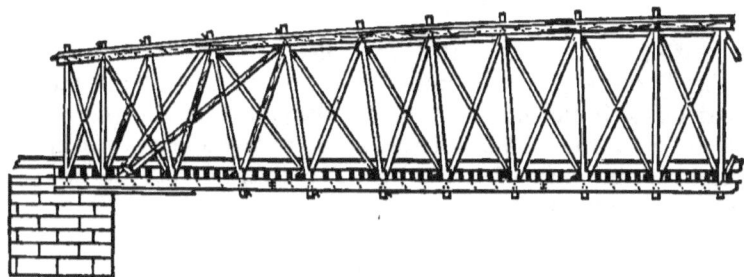

FIG. 102.

braces which pass from the abutment and reach over two or three panels are called *arch-braces*. As the chords are horizontal at the middle, the strain upon them at that point will be the same as for parallel chords throughout, and will be, with sufficient accuracy, $\frac{1}{8}\frac{WL}{D}$, but near the ends the stress will be greater than for parallel chords, because the depth of the truss is less; but this is not a disadvantage, for we have found in our previous investigations that in the latter case the stress is much less than it is at the middle; and when the chords are made of uniform size throughout this is a decided gain. The

---

* *Appleton's Mech. Mag.*, 1852, p. 73.

stress upon the end braces will be nearly $\frac{1}{2} W \sec \theta$, and hence may be the same as for trusses with parallel chords. The stress upon the diagonals and verticals, excepting those which are near the middle, will generally be somewhat less than upon those in trusses having parallel chords, for the upper chord in a measure acts as a brace.

This truss is not necessarily any *stronger* than a Howe Truss of the same span and depth, but it is *stiffer*.

PARABOLIC ARCHED TRUSS.

147.—NOTATION.—This truss has a horizontal lower tie (chord), and a polygonal upper chord, the vertices of the polygon being in the arc of a parabola, the vertex of the parabola being over the middle of the span, and the curve passing through the ends of the span. The trussing may be quadrangular or triangular.

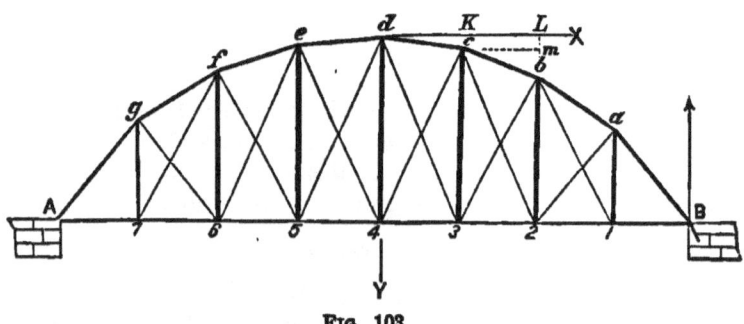

FIG. 103.

In Fig. 103,

Let $N =$ the number of bays in the span;
$n =$ the number of a bay counting from one end;
$c_n =$ the compression on the n-th bay of the upper chord;
$t_n =$ the tension on the n-th bay of the lower chord;
$F =$ the stress on a diagonal;

F_{i} = the stress on a vertical;
i = the inclination of the arch to the horizontal;
θ = the inclination of a diagonal to the vertical;
D = the distance of the vertex of the parabola from the horizontal tie;
h = the depth of the truss at any point;
l = the length of a bay;
p = one of the equal weights placed at the joints; and
V = the reaction of the support at B.

Let the origin of co-ordinates be taken at the vertex of the curve, x horizontal, and y vertical, and
$$x' = d\,K, y' = K\,c,$$
$$x'' = d\,L, y'' = L\,b, \text{ and}$$
$2\,p_{i}$ = the parameter of the parabola.

The general equation of the curve is $x^2 = 2\,p_{i}\,y$, and for the point B this becomes $\tfrac{1}{4} N^2 l^2 = 2\,p_{i}\,D$ $\therefore$ $2\,p_{i} = \dfrac{N^2 l^2}{4\,D}$ and the equation of the curve becomes

$$x^2 = \frac{N^2 l^2}{4\,D} y \text{ or } y = \frac{4\,D}{N^2 l^2} x^2 \ldots\ldots\ldots(172)$$

Counting from B, we have $B\,1$ the first bay; 1–2, the second, and so on to 2–3, which call the n-th. Then
$$x' = (\tfrac{1}{2} N - n)\,l, \text{ and}$$
$$x'' = (\tfrac{1}{2} N - n + 1)\,l, \text{ which}$$
values in Eq. (172) give

$$y' = \frac{D}{N^2} (N - 2\,n)^2, \text{ and}$$

$$y'' = \frac{D}{N^2} (N - 2\,n + 2)^2$$

For the diagonals which incline to the right, as $b\,3$, we have

$$tang\,\theta = \frac{3-2}{b2} = \frac{l}{h} = \frac{l}{D - y''} =$$
$$\frac{N^2\,l}{D\,[N^2 - (N - 2\,n + 2)^2]} \ldots\ldots\ldots\ldots(173)$$

For the diagonal $c\,2$ we have

$$\tan \theta = \frac{3-2}{c\,3} = \frac{l}{D-y'} = \frac{l\,N^2}{4\,n\,D\,(N-n)} \cdot (174)$$

$$\tan i = \frac{y''-y'}{l} = \frac{4\,D}{l\,N^2}(N-2\,n+1)\ldots\ldots(175)$$

As in trusses with parallel chords, the diagonals may be either ties or braces. They are usually ties, and we will treat them as such in the analysis.

148.—CASE OF UNIFORM LOAD.—*Suppose that equal weights are placed at the joints* 1, 2, 3, *etc., throughout.*

Then $V = \tfrac{1}{2}(N-1)\,p$.

If the bay 3–2 be severed, the truss will apparently turn about the joint b. Take b as the origin of moments. The load between B and the n-th bay will be $(n-1)\,p$, and its lever arm will be $(\tfrac{1}{2}n-1)\,l$, and the lever arm of V will be $(n-1)\,l$; and the lever arm of t_n will be $h = D - y''$.

$$\therefore t_n\,h = V(n-1)\,l - (n-1)\,p\,(\tfrac{1}{2}n-1)\,l$$

$$\therefore t_n = \frac{l\,N^2}{8\,D}\,p = \frac{p\,l}{l}\ldots\ldots\ldots\ldots\ldots(176)$$

Hence the stress on the lower chord is uniform throughout for a uniform load.

The same result follows if the load be upon the joints of the upper chord.

The same result also follows if c be taken as the origin of moments.

Hence it is evident that the same result follows if the diagonals are braces instead of ties.

To find the stress upon the diagonal **TIES** *for equal weights placed at all the lower joints.*

It is evident that the horizontal component of the compression on the upper chord *minus* the horizontal component of tension of the diagonal (tie, $b\,3$) must equal the tension on the horizontal tie BA.

$$\therefore c_n \cos i - F \sin \theta = t_n = \frac{l\,N^2}{8\,D}\,p\ldots\ldots(177)$$

The sum of the vertical components of the same strains equals the vertical shearing stress, or

$$c_n \sin i + F \cos \theta = V - \Sigma p = \tfrac{1}{2}(N - 2n + 1)p \ldots (178)$$

Eliminating c_n between Equations (177) and (178), and we have

$$F \cos \theta = \frac{\tfrac{1}{2}(N - 2n + 1) - \dfrac{l N^2}{8 D} \tan \theta}{1 + \tan \theta \tan i} p.$$

Substitute in the second member of this Equation the values of $\tan \theta$ and $\tan i$, Equations (173) and (175), and reducing gives

$$F \cos \theta = 0,$$

hence, *there will be no stress on the diagonal ties for equal weights placed at all the joints of the lower chord.*

The same result is true if the diagonals are braces.

It is also evident that there will be no stress on the diagonals for equal weights on all the joints of the upper chord.

It also becomes evident that for equal weights on all the joints of the lower, the stress on the verticals is equal to p.

It is also evident that for equal weights on all the joints of the upper chord that there will be no stress on either the diagonals or the verticals; and in such a case their only office will be to support the lower chord.

Making $F = 0$ in Eq. (177) and we have

$$c_n \cos i = t_n \ldots \ldots \ldots \ldots \ldots (179)$$

or the horizontal component of stress on the upper chord equals the stress on the lower chord; and hence it is least at the middle and greatest at the ends.

149.—CASE OF A PARTIAL UNIFORM LOAD.—Calling 3-2 the n-th bay, as before, let the load extend from 3 to A, and be removed from 2 to B.

We shall have

$$V = \frac{(N - n)(N - n + 1)}{2 N} p$$

$$t_n = \frac{V(n - 1) l}{D - y''} = \frac{V l N^2}{4 D (N - n + 1)}$$

$$c_n \sin i + F \cos \theta = V \brace c_n \cos i - F \sin \theta = t_n \quad \ldots\ldots\ldots(180)$$

$$\therefore F \cos \theta = \frac{V - t_n \, tang \, i}{1 + tang \, \theta \, tang \, i} \ldots\ldots(181)$$

Substituting $tang \, \theta$ Eq. (173), and $tang \, i$ Eq. (175), and we have

$$F \cos \theta = V \frac{1 - \dfrac{n-1}{D-y''} l \times \dfrac{4 D}{N^2 l}(N - 2n + 1)}{1 + \dfrac{l}{D-y''} \times \dfrac{4 D}{N^2 l}(N - 2n + 1)}$$

$$\therefore F = \frac{(N - n + 1)(n - 1)}{2 N} \frac{p}{\cos \theta} \ldots\ldots (182)$$

which will be the stress on the tie over the n-th bay. n may have all values from $n = 2$ to $n = N - 1$, and hence the expression never becomes negative. It will be necessary, therefore, in order to provide for loads moving in opposite directions, to have diagonals incline both ways in all the panels except the end ones.

If we make $n - 1 = n_1$, the preceding Equation becomes

$$F_{n_1+1} = \frac{(N - n_1) n_1}{2 N} \frac{p}{\cos \theta} \ldots\ldots\ldots\ldots (183)$$

which is a more simple expression, and gives the stress on the tie over the $(n_1 - 1)^{th}$ bay. n_1 may have all values from $n_1 = 3$ to $n_1 = N$.

The stress on the $(n - 1)^{th}$ vertical will be

$$F_{n-1} = \frac{(N - n + 1)(n - 1)}{2 N} p \ldots\ldots \ldots\ldots (184.)$$

It is unnecessary to consider the stress on the upper and lower chords for this case, as it is evident that it will be greatest on those members when all the joints are loaded.

150.—CASE OF DIAGONAL BRACES.—It evidently makes some difference whether the diagonals are *braces* or *ties* when the truss is partially loaded. Thus, when the joints 3, 4, 5, etc., to 7, only are loaded, Fig. 103, if the diagonals are ties, the member *b* 3 will be the active one, but if the diagonals are *braces*, *c* 2 will be the active member for the same load.

PARABOLIC ARCHED TRUSS.

When the diagonals are braces, we may find that the stress on the brace over the n-*th bay is*

$$F = \frac{(N-n)}{2N} \frac{n \; p}{\cos \theta}. \quad \ldots \ldots \ldots \ldots \ldots \ldots \ldots (185)$$

in which n may have all values from $n = 2$ to $n = N - 1$

EXAMPLE.—Let $N = 8$, $D = 2l$.

The value of θ found from Eq. (174), for each particular brace, and substituted in Eq. (185), gives the following results:—

No. of the brace, or, $n =$	Inclination of the braces, or θ Eq. (174).	Values of $\cos \theta$.	Stress on the brace over the n-th bay or F, Eq. (185).
2 or $b1$	33° 41′	0.6587	0.9013 p
3 or $c2$	28° 4′	0.8824	1.0624 p
4 or $d3$	26° 34′	0.8944	1.1182 p
5 or $e4$	28° 4′	0.8824	1.0624 p
6 or $f5$	33° 41′	0.8321	0.9013 p
7 or $g6$	48° 48′	0.6587	0.6641 p

We see here that the diagonals equally distant from the centre are equally strained; and hence, the maximum stress on diagonal braces in any panel will be nearly the same whether the longer or shorter segment is loaded. A close inspection of the figure, in connection with results as found from the solution of an example, will show that they will not be exactly the same.

The strains on verticals, $F_1 = \dfrac{(N-n)n}{2N} \, p$, for a partial uniform load will

be less than p when $N =$ or < 8, in which case the strains on the verticals will be greatest for the load which rests directly upon them. For $N > 8$ the strains due to a partial uniform load will exceed p, on those near the centre.

151.—TRIANGULAR TRUSSING — PARABOLIC ARCHED-TRUSS.—Let the span, Fig. 104, be divided into equal bays, and where verticals through the middle of the bays meet the parabolic arc, will be the vertices

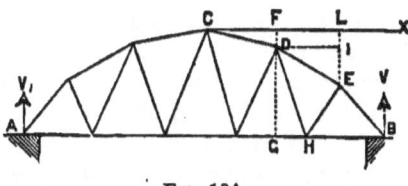

FIG. 104.

of the triangles which form the trussing. The parts between the vertices of the triangles are straight, and form the polygonal upper chord.

It may be shown in this case that when all the joints of the upper chord are loaded, there will be no stress on the triangular trussing, and hence there will be no cross strains on the arch.

Also that when all the joints of the lower chord are loaded, the trussing simply transmits the load to the upper chord, and hence the strains on the diagonal trussing will be $\frac{1}{2} p \sec \theta$ nearly (not exactly, because the tie-braces on each side of the load are not equally inclined).

For a case of partial load, suppose that each of the joints of the lower chord, except H, are loaded with a weight p. The bay at G we will call the n-th. We then have for the tie-brace EH, which may be called the second one of the n-th pair of tie-braces, the following values.*

Inclination of DE to the horizontal $= tang\ i = \dfrac{4\,D}{N^2 l}(N - 2\,n - 2)$

Inclination of EH to the vertical $= tang\ \theta = \dfrac{l\,N^2}{2\,D\,[\,N^2 - (N - 2\,n + 1)\,]^2}$

Reaction of the support at B,

$$V = \frac{(N-n)\,(N-n+1)}{2\,N} p.$$

Stress on $EH = F = \dfrac{(N-n)\,(N-n+1)}{2\,N}\left[\dfrac{4\,n^2 - 1}{4\,n\,N - 4\,n^2 - 1}\right]\dfrac{p}{\cos\theta}$ (186)

PARTIAL LOAD ON THE UPPER CHORD.—If each of the joints from D to A are loaded with a weight p, we may readily find that the stress on EH is

$$F = \frac{(N-n)^2}{2\,N}\left[\frac{4\,n^2 - 1}{4n\,N - 4\,n^2 - 1}\right]\frac{p}{\cos\theta}$$

and the same form of expression is true for DH, but the $\cos\theta$ is not the same for the two tie-braces.

This expression may be put under the form

$$F = \left[\frac{(N-n)\,n}{2\,N} - \frac{(N-n)\,(N-2\,n)}{4\,(N-n)\,n - 1}\right]\frac{p}{\cos\theta}$$

an examination of which shows that the tie-braces are strained more when the short segment is loaded than when the long one is loaded.

152.—STRAINS ON THE UPPER CHORD — FOUND BY MOMENTS.—The strains on the upper chord may be found by solving Equations (180), observing to have the proper value in the second member of the first of those equations for the particular case. It is correct now for a certain partial load; but if the load extends throughout, whether uniform or not, the Equation becomes

$$c_n \sin i + F \cos \theta = V - \Sigma p.$$

---

* See *Jour. Frank. Inst.*. Vol. XLVII., 3d Series, 1864, p. 228.

But it is generally preferable to find the value of c_n directly by moments. Thus, in Fig. 103, to find the stress on cb, if that bay be severed, and the diagonals are ties, the truss would fail by turning about the joint 3. Take 3 as the origin of moments, and let fall a perpendicular from 3 on to cb, and call its length $h' = h \cos i$ (h being the depth $c\,3$). Then, by the principle of moments, we have

$$c_n \times h' = V \times B3 - p_1 \times (3-1) - p_2 \times (3-2)$$

in which p_1 is the load at 1, and p_2 the load at 2.

153.—A GENERAL PROBLEM.—REQUIRED THE FORM OF THE UPPER CHORD SO THAT THE STRAINS ON IT SHALL BE UNIFORM THROUGHOUT FOR A UNIFORM LOAD EXTENDING OVER THE SPAN.

We have seen, when the chords are horizontal, that the greatest stress is at the middle of the span, and when the lower chord is horizontal and the upper chord is a parabolic arch that the strains are greatest near the ends. May there not be some form such that the strains shall be uniform throughout ? *

Let $L = AB = $ the length of the span;
$d = OP = $ the lever arm of the stress on any bay of the upper chord;
$d_1 = CE = $ the value of d at the centre;
$N = $ the number of bays in the span;
$l = $ the length of each bay.

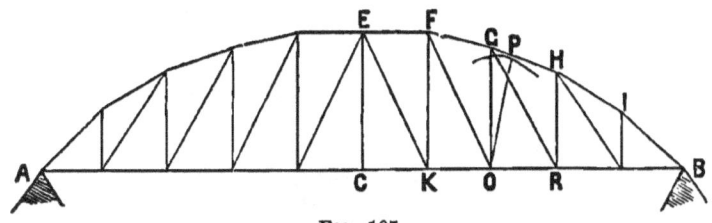

FIG. 105.

$n = $ the number of the bay counting from either end;
$m = $ the number of the central bay ($\frac{1}{2}N$ if N is even);
$c = $ the compression on the upper chord;
$p = $ the load at each joint.

---

* "Solution by Moments," by W. C. Willits, class of 1870, *Univ. of Mich.*

Let the bay GH be severed,—take the origin of moments at O, and we have
$$cd = \tfrac{1}{2}(N-1)\,p\,n\,l - \tfrac{1}{2}(n-1)\,l\,n\,p$$
$$= \tfrac{1}{2}(N-n)\,n\,p\,l$$
For the middle of the span, this Equation becomes
$$cd_1 = \tfrac{1}{2}(N-m)\,m\,p\,l$$
Eliminating c and solving gives
$$d = \frac{(N-n)\,nd_1}{(N-m)\,m} \quad \ldots\ldots\ldots\ldots\ldots(187)$$

Assume N, d_1, and L; hence l becomes known. Find the successive values of d in Eq. (187). Then construct the polygon as follows. Divide the span AB, Fig. 105, into parts each equal to l. Begin at the middle C, and erect a perpendicular equal to d_1. Let the diagonals be braces; then will EK be in action and EF will be parallel to AB. Erect the vertical KF, and draw FO. With O as a centre, and a radius OP equal to the computed value of d (which in this case is for $n = 4$ Eq. (187),) describe an arc GP; and through F draw a tangent to said arc, and limit it by a vertical through O. Draw the brace GR and proceed as before to I. The stress on IB may not be the same as from I to E.

In this case the strains on the lower chord will be greater near the ends than at the middle. *Query.*—Can the parts be so arranged that the strains on upper and lower chords shall both be uniform throughout? Can this be done when one is horizontal?

154.—BOTH CHORDS CURVED.—If both chords are curved

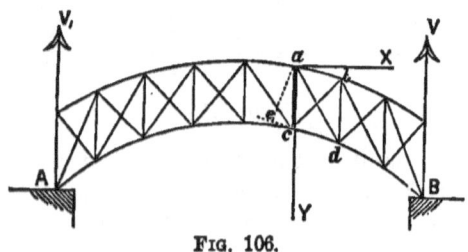

Fig. 106.

upwards, as in Fig. 106, we may find the stresses upon the

chords by the principle of moments; thus, the stress on the bay cd may be found by taking the origin of moments at a. Draw ae perpendicular to the direction of dc, and we have

$$t_n \times ae = V \times ax - \Sigma\, px$$

(ax being the horizontal distance of a from V, and $\Sigma\, px$ a general expression for the moment of all the load between a and B). In a similar way find the stress on $ab = c_n$. Then in Equation (181) substitute $V - \Sigma\, p$ if necessary for V. After this change is made Equation (181) will give the stress on the diagonals, by substituting the proper values for θ and i.

In this way almost any simple truss may be solved, but it is not easy to give general formulas which will facilitate the solution.

CHAPTER V.

COMPOUND STRUCTURES.

155.—REMARK.—By a compound structure I mean the combination of two or more simple structures. The Bollman Truss, as usually made, Figs. 47 and 48, is a compound structure. The main system is a succession of king-posts, but the secondary system is a panel structure. The more common compound systems are made by combining a simple arch with some form of trussing. The compound systems do not admit of as thorough and exact analysis as the simple systems, for they act on different principles, and it is impossible to tell just how much each sustains. The more noted ones are wooden structures, and as the material yields in one system it throws the strain upon the other, and thus there is a constant tendency to equilibrium. Such systems are usually made excessively strong, so that one acts as a safeguard to the other. The wooden structures that have been made on this plan are doubtless much *stiffer* than if all the material were put into either of the simple systems. As the arch plays an important part in these structures, I will here briefly state that if the arch is so made or held that it will not distort, we may find the crushing force at the crown by the formula,

$$H = \frac{WL}{8D} \dots \dots \dots \dots \dots \dots (188)$$

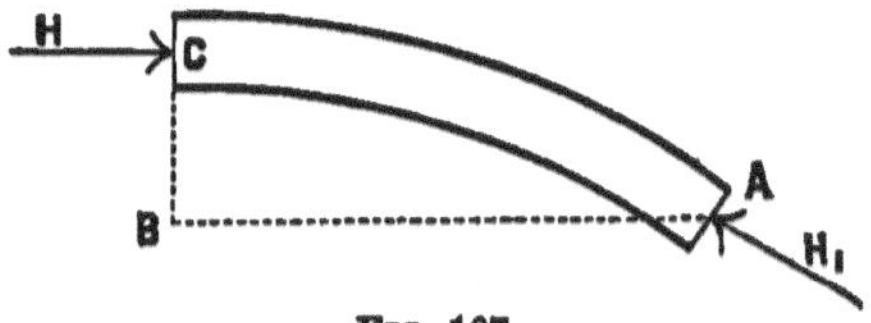

FIG. 107.

in which $L =$ the span $= 2\,AB$;
$D =$ the versed-sine $= CB$;
$W =$ the load on the span; and
$H =$ the thrust at the crown.

This formula is only true if the arch receives its thrust at the abutment, but if it is supported at its ends it becomes simply a curved beam, and there will be both tension and compression at C. As these arches are usually very flat, the compression at the ends will rarely exceed more than 10 per cent. above that at the crown.

156.—BURR TRUSS.—The Burr Truss, Fig. 108, is a wooden structure and was in very common use in some parts of the country not many years since. The arrangement is so evident

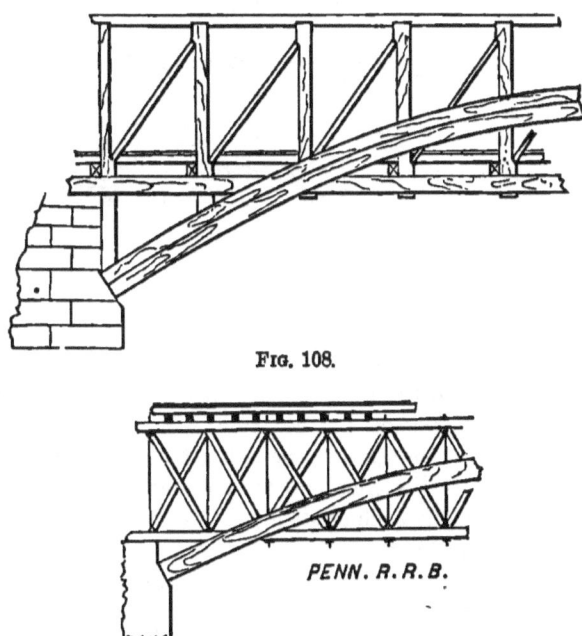

FIG. 108.

PENN. R.R.B.

FIG. 109.

from the figure that a special description is deemed unnecessary. It appears that the arch was relied upon to resist the strains which would, in the panel system, fall upon the counter-braces.

157.—SECOND EXAMPLE.—Fig. 109 is another example of a compound structure which is composed of a Howe Truss and a simple arch. It is evident from the remarks previously made, that there is an unnecessary number of counter-braces. The arch is usually secured to the truss by bolts, where the members cross each other.

158.—THIRD EXAMPLE.—Fig. 110 is another example of a compound structure. In this the truss is, in a measure, suspended from the arch by suspension rods. There are no counter-braces. Instead of counter-braces in the panels, there are iron ties, which incline the same way as the braces, and hence serve as counter-braces when necessary. It is very doubtful if

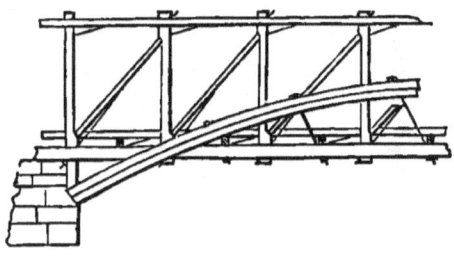

FIG. 110.

any are needed near the ends. When either end half of the arch is loaded the other half tends to thrust upward, which tendency should be resisted by the truss; hence the necessity of counter-braces, or counter-ties near the middle; but near the ends the weight of the truss and arch would probably be sufficient to overcome any such tendency. On this point see the discussion on counter-braces.

Numerous other forms might be given, but these will answer to illustrate the topic.

To analyze these structures, ascertain the load which the arch will carry, and then the load which the truss will carry, and add the results. *The margin of safety* for both combined should considerably exceed that which would be allowed if either, acting separately, carried the whole load, for reasons previously given.

Part 3.

ROOFS.

159.—A ROOF, in common language, is the covering over a structure, the chief object of which is to protect the building against the effects of snow and rain. It is composed of boards, shingles, slate, mastic, or other suitable materials.

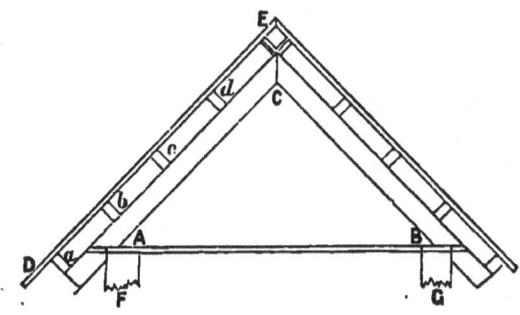

FIG. 111.

The inclined pieces AC, and BC, Fig. 111, which support the roof are called RAFTERS. When the roof is light, the roof boards DE are placed directly upon the rafters, but when the rafters are far apart, say more than four feet, small pieces a, b, c, and d, called PURLINES,* are placed across the rafters for the purpose of receiving the roof proper. AB is a tie, and F and G represent the ends of posts. The frame ABC is called a *roof truss*.

160.—ROOF TRUSSES have a great variety of forms, and differ greatly in the details of their construction. All the trusses which have been discussed in the preceding pages are

---
* Purline *beams* are sometimes placed under the rafters.

suitable for this purpose in many cases. Some other forms are given in the following pages.

161.—GENERAL DATA.—A roof truss is required to carry its own weight, the weight of the purlines, the weight of the roof above them, the force of the wind, the weight of snow when there is any, and in some cases certain local or concentrated loads, such as floors, machinery, and the like, which are suspended from the roof trusses.

162.—DESCRIPTION OF THE ROOF OVER THE LARGE HALL OF THE UNIVERSITY OF MICHIGAN.—The roof over the large hall of the University of Michigan contains some novel features, and in some respects is a bold design. The outline and arrangement of the parts, including the dome, were designed by the architect,* but the details of the large trusses were arranged and proportioned by the author, and erected under his superintendence; hence they possess a peculiar interest to him. It is here presented as a practical problem.

The frontispiece shows a vertical section from east to west of the dome and roof, through the centre of the dome, excepting that the truss is shown as if it was between the eye and the dome. The west end apparently rests directly upon the columns which support the roof, but in reality nearly the whole dome rests upon the trusses. A skeleton of the elevation and plan of the truss is shown in Fig. 112, and a plan of the roof in Fig. 113. These Figs. are not drawn to the same scale.

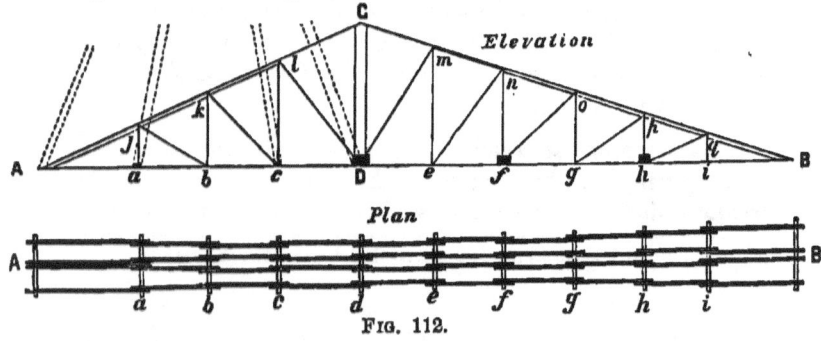

Fig. 112.

---

* The New University Hall was designed by Mr. E. S. Jenison, a student of the author, class of 1868, *Univ. of Mich.*

ROOF TRUSS. 195

The width *xy*, Fig. 113, over which the trusses are placed, is 80 feet, and the length *wy* is 128 feet 10 inches. There are two trusses, each like that shown in Fig. 112, placed across the space, one at *AB*, and the other at *A'B'*, the distance between them from centre to centre, being 34 feet. Side trusses, of the

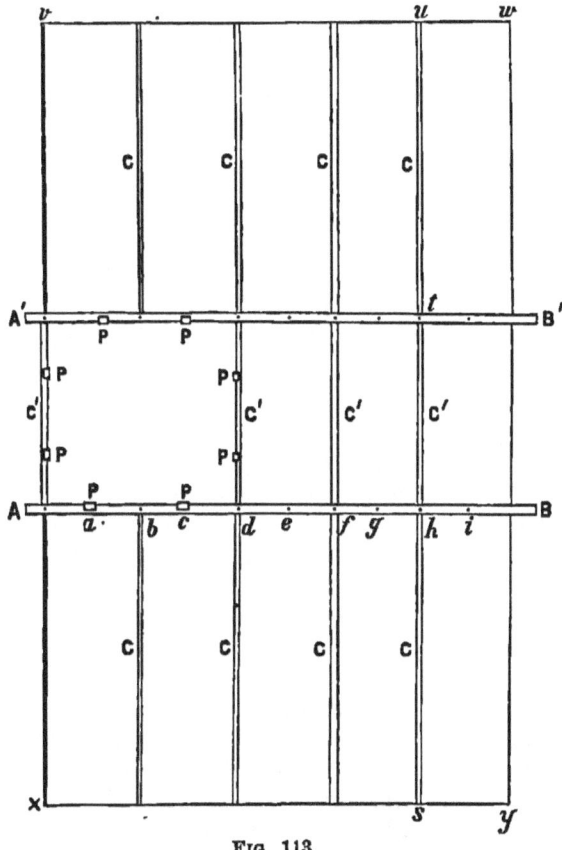

FIG. 113.

King-Post style, extend from the main trusses to the end walls, and are marked *C, C, C*, in Fig. 113. These are for carrying that portion of the roof which is between the main trusses and the end walls. Trusses of a similar style *C', C', C'*, were placed between the main trusses.

The dome is built around and secured in place by eight long posts, twelve inches square, made of joists which are two inches thick by twelve inches wide, breaking joints, and firmly bolted together. The lower ends of these posts rest upon and are secured to trusses at P, P, P, Fig. 113. Four of these are secured directly to the main trusses, and the other four to cross trusses. The cross truss under the right hand (or east) side of the dome is 33 feet from the west end of the main truss. The posts all meet each other at a common point near the upper end of the dome, as shown in the frontispiece, and are firmly secured to each other; hence, they may be considered as forming the edges of a regular octagonal pyramid. The height of the dome above the lower side of the main trusses is seventy feet, and the base covers about 50 feet square on the main roof.

Purlines, which were two inches thick by twelve inches deep, were placed upon the cross (or side) trusses C, C, C'', C'', and the whole was covered with a continuous flat roof, having a pitch of one-half of an inch to the foot, thus forming a large base—120 by 128 feet, including the roof in the rear of the main hall —for resisting the force of the wind on the dome. All the trusses—the main and cross trusses—extend downward into the roof, and were made use of for dividing the ceiling into panels. The panels were over four feet deep and extended from the walls to the main truss on the sides, and from one main truss to the other in the middle, the panels under the main trusses crossing the others at right angles. The position of d, Fig. 113, was determined by the size of the base of the dome. The side truss at b is midway between A and d, and those at f and h divide the space dB into three equal parts.

The main rafters AC and CB, Fig. 112, are solid pieces of pine, fourteen inches wide and sixteen inches deep. The upper part of the rafter CB is above the roof and cased in. The truss ACB is called the **PRIMARY TRUSS**. The rafters of the **SECONDARY TRUSS** are formed of several pieces, each one of which extends between two consecutive joints,—as Aj, jk, kl, mn, etc. These are bolted to the main rafters. The pieces jb, kc, lD, Dm, etc., are braces of the secondary truss.

The tie AB is common both to the main and secondary trusses. It is composed of flat bars (or links) of iron, of uniform thickness and of different depths, depending upon the amount of stress to which they are subjected. They are enlarged at the end, and have an eye for receiving a pin, as shown in Fig. 113a. Four of these are placed side by side, and arranged as shown in the plan in Fig. 112. Cast-iron blocks rest against the pins for receiving the end of the braces of the secondary trusses.

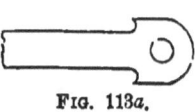

FIG. 113a.

At each end of the trusses at A and B is a large cast-iron block, which weighs 680 lbs. These are for receiving the pressure of the main and secondary rafters. The bars of the main tie pass through two slots in these pieces, and are secured on the outside of the blocks by a large pin $3\frac{1}{2}$ inches in diameter.

The vertical members aj, bk, etc., are iron ties, the lower ends of which pass through holes in a cast-iron block, Fig. 114a, and secured by nuts on the under side. This block is placed below the large tie. The upper ends are secured in a similar way. In some cases there are two and in other cases three ties, which are represented in Fig. 112 by a single line, the number and position of which will be given hereafter.

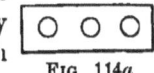

FIG. 114a.

As a general thing, one or two rods, as the case might be, was used for supporting the load which was placed at a joint, and the other rod at that joint was used for supporting the transmitted pressures, although this was not always the case. The conditions will be fully shown in the following analysis.

163.—THE LOAD ON THE FLAT PART OF THE ROOF.— It is not supposed that all the data which are given below are applicable to all cases, or even that there are no questions in regard to their correctness. They are presented as they were used at the time, and are open for discussion. It was intended, however, to be on the safe side in all cases, and it is advisable, in a case like this, where the ceiling is secured directly to the roof, and the roof acted upon by a large dome under the varying pressures of the wind, to be largely on the safe side, so as to avoid as far as possible breaks and cracks in the ceiling.

I supposed that the several loads were reduced to an

equivalent load of pine uniformly distributed over the whole roof.

Inches thick
of pine.

Felt and mastic roof........................... = 1.50*
Roof boards.................................. = 1.00
Roof joists (2″ × 12″, — 16″ apart)............ = 1.50
Lath joists (2″ × 6″, — 16″ apart)............. = 0.75
Panel-work in the ceiling..................... = 0.75
Side truss, tie-rods, etc...................... = 2.00
Plaster, including panels, cornices, etc........ = 3.00
Weight of snow and pressure of the wind........ = 6.00

Total..........16.50

The weight of four specimens of pine, taken at random, were,

1st specimen 32 lbs. per cubic foot.
2d specimen 40 lbs. per cubic foot.
3d specimen 30 lbs. per cubic foot.
4th specimen 45 lbs. per cubic foot.

Total..... 147 lbs.

Average 37 lbs. per cubic foot.

Two of the specimens appeared to be quite thoroughly seasoned, and the other two only partially so. I assumed 37 lbs. as the weight per cubic foot of the material, which must have been on the safe side. This gives, according to the preceding data (16½ × 37 ÷ 12 =), 54 lbs. (nearly) for the load upon a square foot of the roof.

Francis, in his book on "Iron Columns," p. 19, says that flat roofs loaded with snow are liable to weigh 50 lbs. per square foot; and Trautwine, in his "Engineer's Pocket-Book," p. 301, says, that when the roof is plastered below, the weight may be 46 lbs. per square foot, including the weight of the snow and pressure of the wind. As these authors could not have had in mind such heavy panel-work in the ceiling as exists in

---

* This was the estimate of the roofer, but I now think it should have been 2¼ to 3 inches. Within two years after the erection of the building, the mastic was removed, and the roof was covered with tin.

this case, and as I wished to keep decidedly on the safe side, I adhered to the above result of 54 lbs.

This gives for the weight of the roof, including snow, etc., and excluding the weight of the dome, and the space occupied by the dome, 500,000 lbs., and for the weight when the timber is thoroughly seasoned (called 28 lbs. per cubic foot) and exclusive of the weight of the snow, 193,000 lbs.

The weight of snow and the pressure of wind are deserving of special notice.

164.—THE WEIGHT OF SNOW.—Freshly fallen snow weighs from five to twelve lbs. per cubic foot, although snow which is saturated with water weighs much more. Very wet snow rarely falls to a very great depth, especially in the southern part of Michigan. Some say that snow is equivalent to from $\frac{1}{10}$ to $\frac{1}{8}$ of its depth in water, while others say that it may be equivalent to $\frac{1}{4}$ its depth of water.

European engineers consider that six lbs. per square foot is sufficient for snow, and eight lbs. for the pressure of the wind, making fourteen lbs. for both. Trautwine says that not less than twenty lbs. should be allowed in the United States.

As the roof in the case which we are considering is flat, a large quantity of snow may rest upon it, but the pressure of the wind upon it will probably be small.

Snow in the vicinity of Ann Arbor is rarely three feet deep on the ground, but because it is sometimes that depth, or of an equivalent depth of heavy snow, the load due to its weight must be provided for. Calling the snow equivalent to $3\frac{1}{2}$ inches deep of water, and we find that it equals 6 inches nearly of partly seasoned pine—which is $18\frac{1}{2}$ lbs. per square foot —a value somewhat under that assumed by Mr. Trautwine, but which I consider quite large enough in this case. Before so much snow can fall upon the roof, all the timbers in it will be lighter from seasoning than that assumed above, so that the entire roof will be lighter than that which we have assumed.

165.—THE FORCE OF THE WIND.—The pressure of the wind upon the dome is of special importance in considering its stability and in proportioning the trusses. According to Mr.

Smeaton, the pressure of the wind directly against a flat surface in a hurricane may be 32 lbs. per square foot. Tredgold recommends the use of 40 lbs. per square foot. A gauge in Girard College broke under a strain of 42 lbs. per square foot, whilst a tornado was passing near by. During the severest gale on record at Liverpool, England, there was a pressure of 42 lbs. per square foot directly upon a flat surface. During a very violent gale in Scotland, a wind-gauge once indicated 45 lbs. per square foot. Buildings which are more or less protected will not be subjected to such pressures.

Although there are high winds at Ann Arbor, yet no such gales as those mentioned above have ever been known there; at least I judge so from the fact that comparatively little damage has been done by high winds. But if such winds do occur it will be safe to assume less than 40 lbs. per square foot on account of the oval shape of the dome;—and also because materials will sustain a high strain for a short time without apparent damage. If, therefore, we should proportion the parts for 25 lbs. pressure, they would doubtless sustain 50 lbs. without damaging them; and to avoid much deflection in the trusses in case of a strong wind it is not advisable to use a smaller value. I therefore used 25 lbs. per square foot upon a meridian section of the dome.

The cylindrical part of the base of the dome is about 34 feet in diameter, and the total height above the angle of the truss is about 55 feet. To get the force of the wind I called the average diameter 22 feet, and 60 feet high. This gives a pressure of 33,000 lbs.

The centre of pressure is about 30 or 32 feet above the fastenings of the lower ends of the posts of the dome, and hence the pressure on the feet of the posts due to the pressure of the wind will nearly equal the pressure of the wind upon the surface of the dome. The wind may blow from any direction. When it is directly in the north or south it will tend to throw the dome directly upon one or the other of the large trusses, and I assumed that each of the two posts which rest on the truss would carry 15,000 lbs. due to the pressure of the wind; and when the wind is in the east or west it may cause the same pressure on the posts on the opposite side.

But to be completely on the safe side in proportioning the main trusses, I assumed that, when the wind is in the north-west, or south-west, the whole of the 30,000 lbs. would be thrown directly upon the angle of the large truss. This being done, it is unnecessary to consider the transmitted strain of 15,000 lbs. which is thrown upon the other points when the wind is in other directions, as will be noticed in the analysis.

166.—WEIGHT OF THE DOME.—As it was difficult to determine the exact weight of the dome from the drawings, I made a rough estimate of it by assuming that it was equivalent to a hollow cylinder whose mean diameter was 24 feet, height 80 feet, and whose thickness was six inches of solid pine. This gave a weight of 112,000 lbs. The architect computed from a bill of materials that it would weigh 96,000 lbs., but I do not know what he used for the weight of pine. A review since its erection gave 107,000 lbs., at 37 lbs. per cubic foot.

There were eight supports, as before described, each of which I assumed would sustain one-eighth of the load, or 14,000 lbs. each. In reality these posts did not sustain this amount, for a large portion of the dome rested directly upon the large trusses, but the computation would be essentially the same, excepting that in the latter case it would make the strain upon some of the vertical tie-rods less than that found by the following computation.

167.—WEIGHT OF THE MAIN TRUSSES.—From the bill of materials I found that the large trusses would weigh 17,800 lbs. each; and hence at each of the joints a, b, c, etc., Fig. 112, the weight will be 1,780 lbs., assuming that the load is uniformly distributed and that each joint carries one-tenth of it.

168.—WEIGHT OF CROSS TRUSS.—There was a truss between the two large trusses at D, to keep them erect, the weight of which is 3,600 lbs., or 1,800 lbs. on each truss.

169.—RESULTS COLLECTED.—At a, Fig. 112, the load is one eighth of the weight of the dome, or 14,000 lbs.; the pressure due to the wind, or 15,000 lbs.; *plus* one-tenth of the weight of the truss, or 1,800 lbs.

At b the load is one-half the weight on the side truss due to the weight of the roof, or 11,780 lbs.; *plus* that due to the weight of the snow, or 6,150 lbs.; *plus* one-tenth of the weight of the large truss, or 1,800 lbs.; which together equals 19,730 lbs.

At D the load is one-half the load on one side truss, *plus* one-half that on the cross truss C' (which supports one-half the load between $C'd$ and $C'f$), or 12,690 lbs.; *plus* the load due to snow, or 7,880 lbs.; *plus* one-tenth of the weight of the truss, or 1,800 lbs.; *plus* one-eighth of the weight of the dome, or 14,000 lbs.; *plus* the pressure due to the wind, or 30,000 lbs.; *plus* one-half the weight of the cross truss, or 1,800 lbs.

Similarly, we find the loads on e, f, g, etc. These results are brought together in the following tabular form:—

Weights in lbs. due to the several loads on the several joints.

Weight	a	b	c	D	e	f	g	h	i	Support A
Roof		11,780		12,690		18,275		18,275		27,000
Snow		6,150		7,880		9,625		9,625		14,000
Truss	1,800	1,800	1,800	1,800	1,800	1,800	1,800	1,800	1,800	9,000
Dome	14,000			14,000						44,000
Pressure due to wind	15,000			15,000	30,000					30,000
Weight of cross truss					1,800					2,000
Totals	30,800	19,730	30,800	63,170	1,800	29,700	1,800	29,700 900	1,800	126,000
								30,600		

The load at i being small, the brace hg was omitted, and in the analysis the load at h was called $29,700 + 900 = 30,600$ lbs.

The load on the support at A was used for determining the dimensions of the column which supported one end of the main truss.

170.—ANALYSIS.—We have the following dimensions:—

$AB = 80$ feet;
$CD = 16$ feet;
$Bi = 7$ feet 10 inches $= ih = hg = gf = fe = eD$;
$aA = 8$ feet;
$ab = 8$ feet 6 inches $= bc = cD$;
$AD = 33$ feet, and
$DB = 47$ feet.

By means of these quantities we can find the lengths of the verticals and of the inclined braces, and the angles which they make with each other. In regard to the latter a practical difficulty presents itself. For the timbers having finite dimensions it is found practically impossible to put them in place and give them the proper bearing, and at the same time secure the same inclination that we would have if the parts were reduced to mathematical lines.

In this case, if the centre lines of the main rafters AC and BC are prolonged, they will meet the line of the main tie several feet outside of the points of support, and the question arises whether we shall use the angles as they exist in the structure, or rely upon the dimensions above given. Neither will be exactly correct, but it will be nearer correct to use the former than the latter. I have therefore determined the several angles of inclination with the vertical from a scale drawing, and used the secants and tangents to the nearest tenth.

If the load at a is 30,800 lbs., see Article 169, we may assume that it is supported by the secondary truss Ajb, and hence, according to Article 46, we have

stress on aj = 30,800 *lbs.*;
stress on Aj = $\frac{1}{2}$ of 30,800 *lbs.* × *sec* ajA;
stress on bj = $\frac{1}{2}$ of 30,800 *lbs.* × *sec* ajb;
stress on Ab = $\frac{1}{2}$ of 30,800 *lbs.* × *tang* ajb.

The vertical pressure at b is $\frac{1}{2}$ of 30,800 lbs., *plus* the load placed at b, or 19,730 lbs.; hence the total vertical pressure at b is 35,130 lbs. We may assume that this is carried by the secondary king-post truss Akc; and hence we have

stress on bk = 35,130 *lbs.*;
stress on Ak = $\frac{1}{2}$ of 35,130 *lbs.* × *sec* Akb;
stress on ck = $\frac{2}{3}$ of 35,130 *lbs.* × *sec* bkc;
stress on Ac = $\frac{1}{2}$ of 35,130 *lbs.* × *tang* bkA; or
stress on Ac = $\frac{2}{3}$ of 35,130 *lbs.* × *tang* bkc.

Similarly, the vertical pressure at c is $\frac{2}{3}$ of 35,130 lbs. *plus* the load, 30,800 lbs. at c; hence, the total load is 54,220 lbs. This is supported by the secondary truss AlD. Hence, we have

stress on cl = 54,220 *lbs.*;
stress on Al = $\frac{1}{4}$ of 54,220 *lbs.* × sec clA;
stress on Dl = $\frac{3}{4}$ of 54,220 *lbs.* × sec clD;
stress on AD = $\frac{3}{4}$ of 54,220 *lbs.* × tang clD.

Hence the reaction at D of the truss AlD is $\frac{3}{4}$ of 54,220 lbs. = 40,665 lbs. This may be proved by finding the reaction directly, thus :—

Stress at D = $\frac{1}{4}$ of the load at a + $\frac{3}{4}$ that at b + $\frac{3}{4}$ that at c; or
= $\frac{1}{4}$ of 30,800 lbs. + $\frac{3}{4}$ of 19,730 lbs. + $\frac{3}{4}$ of 30,800 lbs.;
= 40,665 lbs., as before.

Proceed in a similar way with the secondary trussing between D and B, and we finally find that the point D sustains 34,400 lbs. of the load on DB.

The total *vertical* stress at D is that which is transmitted to D through lD from the left (40,665 lbs.); *plus* that transmitted to D through mD from the right (32,400); *plus* the load which is directly applied at D (68,170); or 141,235 lbs. But we may deduct from this all that part of the 15,000 lbs. which is applied at a and c, due to the pressure of the wind; for these pressures cannot exist at the same time that the 30,000 lbs. does, which is supposed to be applied at D. Hence we may deduct $\frac{1}{4}$ of 15,000 lbs. + $\frac{3}{4}$ of 15,000 lbs., or 15,000 lbs. This leaves for the effective vertical pressure at D, 126,235 lbs. This is supported by the main truss ACB. Hence, we have (omitting all below 100 lbs.),

stress on DC = 126,200 *lbs.*;
stress on AC = $\frac{47}{80}$ of 126 200 *lbs.* × sec DCA;
stress on BC = $\frac{33}{80}$ of 126,200 *lbs.* × sec DCB;
stress on AB = $\frac{33}{80}$ of 126,200 *lbs.* × tang DCB.

Where two or more stresses are common to a single piece, the resultant stress is the sum of all the partial stresses. Thus the *total* stress on Ab, for instance, is the stress on ab of the truss Ajb; *plus* the stress on Ac of the truss Akc; *plus* the stress on AD due to the truss ACB.

Similarly, the total stress on Aj equals the stress on Aj of the truss Ajb; *plus* the stress on Ak of the truss Akc + etc.

As a check upon the work, we have the total stress on Aj equal to the total reaction on the support A multiplied by the secant of ACD; which should equal the sum of all the partial stresses.

(NOTE.—This check would be exact for a skeleton or line truss, but we do not expect it to be exactly true in this case, where the angles are not determined from the dimensions of the truss.)

In ordinary cases it would probably be thought advisable to use 600 or 800 lbs. per square inch for the safe resistance to crushing of pine; but as the maximum load which we have assumed—that due to green lumber, green mortar, a deep snow, and a hurricane, all applied at the same time—will probably never exist, we may safely assume 1,000 lbs.; for it probably might be strained to double this amount for a short time without endangering its strength. This is the value which we have used. The iron will safely resist 12,000 lbs. per square inch. It was all tested to 15,000 lbs. Using these values, and we have the following numerical results. All strains less than 100 lbs. are omitted in the final result. The number and size of the pieces which were used to resist the strain are also given.

VERTICAL TIE RODS.

Name of the Piece.	Total Load.	Suspended Load.	Transmitted Stress.	No. of Iron rods.	Diameter of each rod.
Pieces.	Lbs.	Lbs.	Lbs.		Inches.
aj	30,800	30,800	..	1	$1\frac{3}{4}$
bk	35,100	19,700	..	1	$1\frac{1}{8}$
			15,400	1	$1\frac{1}{8}$
cl	54,200	30,800	..	1	$1\frac{3}{4}$
			23,400	1	$1\frac{3}{4}$
DC	126,265	68,200	..	2	$1\frac{7}{8}$
			58,000	2	$1\frac{7}{8}$
em	38,900	1,800	37,100	2	$1\frac{5}{8}$
fn	46,300	29,700	..	2	$1\frac{3}{8}$
			16,600	1	$1\frac{5}{8}$
go	22,200	1,800	20,400	1	$1\frac{5}{8}$
hp	30,600	30,600	..	2	$1\frac{5}{8}$

THE HORIZONTAL TIE.

Name of the Piece.	Vertical Components of the Forces which Produce Tension.	Tangent of the Angle of Inclination of the Piece which Produces Tension.	Resultant Tension.	No. of Pieces side by side.	Thickness of each Piece, Inches.	Depth of each Piece, Inches.
at D	$\frac{17}{10}$ of 126,200	× 2.4	= 177,900			
or	$\frac{22}{10}$ of 126,200	× 2.6	= 176,000*			
cD	$\{\frac{17}{10}$ of 126,200 $\frac{2}{3}$ of 54,200	×2.4$\}$ ×0.7$\}$	= 206,400	4	$1\frac{1}{8}$	$4\frac{1}{4}$
bc	$\{\frac{2}{3}$ of 35,100 plus the stress on cD	× 1.1	= 232,100	4	$1\frac{1}{8}$	$4\frac{1}{4}$
Ab	$\{\frac{2}{3}$ of 30,800 plus the stress on bc	× 2.4	= 269,000	4	$1\frac{1}{8}$	$5\frac{1}{4}$
De	$\{\frac{23}{10}$ of 126,200 $\frac{2}{3}$ of 1,800	×3.4$\}$ ×0.6$\}$	= 185,900	4	$1\frac{1}{8}$	4
ef	$\{\frac{2}{3}$ of 46,300 plus the stress on De	× 0.8	= 215,500	4	$1\frac{1}{8}$	$4\frac{1}{4}$
fg	$\{\frac{2}{3}$ of 22,220 plus the stress on ef	× 1.0	= 232,200	4	$1\frac{1}{8}$	$4\frac{1}{4}$
gh	$\{\frac{2}{3}$ of 20,600 plus the stress on fg	× 1.3	= 250,100	4	$1\frac{1}{8}$	$5\frac{1}{4}$
gh	$hi = iB$			4	$1\frac{1}{8}$	$5\frac{1}{4}$

* The difference in the two results is discarded.

WOODEN PIECES.

Name of the Piece.	Vertical Component of the Stress.	Secant of the Angle of Inclination of the Piece.	Resultant Stress.	Thickness of the Pieces.	Width of the Pieces.
	Primary Truss.				
	Lbs.		Lbs.	Inches.	Inches.
AC	$\frac{17}{20}$ of 126,265	× 2.6	= 192,700	14	16
BC	$\frac{33}{40}$ of 126,265	× 3.5	= 182,200	14	16
	Secondary Truss.				
kl	¼ of 54,200	× 2.6	= 35,200	3	12
jk	¼ of 35,130, plus the stress on Al	× 2.6	= 65,700	6	12
Aj	¼ of 30,800, plus the stress on jk	× 2.6	= 105,700	12	14
mn	⅛ of 38,800	× 3.5	= 22,600	2	12
no	⅛ of 38,800 ¼ of 46,350	× 3.5	= 55,100	6	12
op	¼ of 22,220, plus the stress on no	× 3.5	= 74,500	8	12
pB	¼ of 30,000, plus the stress on op	× 3.5	= 110,200	12	14
gp	⅔ of 30,600	× 1.7	= 34,600	6	8
fo	⅔ of 22,200	× 1.4	= 23,300	6	12
en	⅔ of 46,350	× 1.3	= 48,800	6	12
Dm	⅚ of 38,880	× 1.2	= 38,900	6	12
Dl	⅔ of 54,220	× 1.2	= 48,800	6	12
ck	⅔ of 35,130	× 1.5	= 35,100	6	12
bj	½ of 30,800	× 2.4	= 39,900	6	12

Many of the wooden pieces were such as the carpenter had in "stock," and are larger than is necessary, as is shown by a comparison of the strains and size of the pieces. It was assumed, however, that the pieces would not bend. The dimensions of the iron pieces in the large tie were changed very slightly from those given by the analysis.

Checks upon the computation.—The reaction at A, omitting the 15,000 lbs. at each of the points a and c, which is due to the force of the wind, will be

$$\frac{15,800 \times 71\tfrac{3}{4} + 19,730 \times 63\tfrac{1}{2} + 15,800 \times 55\tfrac{1}{4} + 68,170 \times 47 +}{80}$$

$$\frac{1,800 \times 39\tfrac{1}{8} + 29,700 \times 31\tfrac{1}{8} + 1,800 \times 23\tfrac{1}{2} + 30,600 \times 15\tfrac{1}{8}}{80}$$

$= 99,800$ lbs.

The 30,000 lbs. omitted above was included in the preceding analysis, in such a way as to be equivalent to $\tfrac{3}{4}$ of 15,000 lbs. supported at A, plus $\tfrac{1}{4}$ of 15,000 lbs. supported at the same point, or 15,000 lbs. in all. Adding this to the preceding value, and we have 114,800 lbs. This multiplied by the secant of the inclination, 2.6, gives 298,480 lbs. for the total stress on Aj. By the preceding analysis we have the total stress on

$$Aj = 105,700 + 192,700 = 298,400 \text{ lbs.},$$

which is the same as the preceding to within 100 lbs., which difference results from dropping small quantities, and which is of no importance in comparison with the total stress.

Similarly, the stress on Ab will be

$$114,800 \times 2.4 = 275,520.$$

But from the preceding analysis we find that the stress is 269,100. This shows a difference of 6,400 lbs., which would make a difference of only one-half of a square inch of iron in the tie. But the latter result is doubtless nearer correct than the former, for it is dependent upon the inclination of the small braces (as well as the main rafter), and the former were measured as they exist.

171.—CAMBRE OF THE LARGE TRUSSES.—From the time that the trusses are erected, to the time of the completion of the roof and dome, they will continue to settle. The causes of tne settling are the increase of load and the elasticity of the material. The point of greatest deflection will evidently be at the joint D, directly under the angle C. After they are erected they will change their deflection from time to time, as the load due to the pressure of the wind and weight of snow changes. It is desirable to make such a cambre in the lower chord as that it shall never fall below a horizontal. We will therefore assume that the change of deflection from what it was when first erected, is that due to the total load, as given in Article 169.

The deflection will be the result of three causes, each of which may be considered independently.

1st, That due to the direct elongation of DC, which is caused by the stress on the vertical ties at DC, while DB and CB are supposed to remain constant.

2d, That due to the compression of CB, whilst CD and DB remain constant.

3d, That due to the elongation of DB, whilst DC and CB remain constant.

The triangle ADC may be considered, instead of CBD, or the computations on one may be used as a check on the other, but as the dimensions of the two do not differ largely, the results will not differ much.

1st, *The elongation of DC* is caused by the application of 68,000 lbs. at D, *plus* a transmitted stress of 58,000 pounds, or a total of 126,000 lbs. The length of each of the four tie rods is 16 feet, and their diameter is 1¾ inch, and hence the cross section of each is 2.76 square inches. The formula for the elongation is

$$\lambda = \frac{Pl}{EK}$$

Calling $E = 28,000,000$ lbs. and the formula becomes

$$\lambda = \frac{126,000 \times 12 \times 16}{28,000,000 \times 11.04} = 0.078 \text{ of an inch.}$$

2d, *The deflection due to the compression of BC* may be supposed, without sensible error, to follow the same law as the differentials of the quantities. Hence, by differentiating the expression

$$CB^2 = CD^2 + BD^2, \ldots\ldots\ldots\ldots\ldots\ldots\ldots (a)$$

considering BD as constant, we have

$$CB\, d\,(CB) = CD.d\,(CD)\ldots\ldots\ldots\ldots\ldots\ldots(b)$$

The length of $CB = \sqrt{BD^2 + CD^2} = 50$ feet, nearly.

$$d\,(CB) = \frac{Pl}{EK} = \frac{182{,}000 \times 12 \times 50}{1{,}500{,}000 \times 14 \times 16} = 0.325 \text{ of an inch.}$$

Hence, from Eq. (b), we have

$$d\,(CD) = \frac{CB.d\,(CB)}{CD} = \frac{12 \times 50 \times 0.325}{12 \times 16} = 1.015 \text{ inch.}$$

3d, *The deflection due to the elongation of DB* is found in a similar way, by differentiating Eq. (a), considering CB as constant.

$$\therefore CD.d\,(CD) = -\, BD.d\,(BD)\ldots\ldots\ldots\ldots\ldots(c)$$

The differential of BD is

$$d\,(BD) = \lambda = \frac{Pl}{EK} = \frac{185{,}900 \times 12 \times 47}{28{,}000{,}000 \times 4 \times 1\tfrac{1}{2} \times 4} =$$

0.208 of an inch.

Although the tie-rod is not of uniform size throughout its length, yet if its section is proportional to the stress to which it is subjected at its several parts, the elongation due to the several stresses will be the same as if we consider it uniform, and under the action of that stress which corresponds to that section. The negative value shows that CD is *shortened* by depressing C whilst DB is *elongated*, but as D is depressed the same amount that C is, the result is essentially positive for our use. Eq. (c) gives

$$d\,(CD) = \frac{12 \times 47 \times 0.208}{12 \times 16} = 0.611 \text{ of an inch.}$$

Hence the total deflection due to all these causes is

$$0.078 + 1.015 + 0.611 = 1.704 \text{ inches.}$$

This computation makes no allowance for imperfection in the joints. In the construction, the lower tie-rod had a deflection downward of several inches, as can be seen from the frontispiece. (This can be tested on the figure by a straight-edge. But the false work which was made below this for supporting the plaster cornice was cambered upward about three inches. As the roof was constructed before this part of the work was completed, it was probably enough to secure a permanent cambre after the dome was constructed, and all the joints had come fully to bearing.

The change of cambre due to a severe wind of 25 lbs. per square foot cannot exceed that due to a load of 30,000 lbs. placed at D, and hence cannot exceed $\frac{30,000}{126,000}$ of 1.704 = 0.405 of an inch; but usually it will not equal one-fourth of this amount. If the wind should blow directly from the west or east with a force of 25 lbs. per square foot, it will not exceed one-half the above amount, or 0.202 of an inch in each case; or about 0.4 of an inch in both cases; that is, in one case it will be below the normal position, and in the other above it.

The change of cambre will affect the perpendicularity of the dome; and as the height of the dome is almost exactly twice the width of the space between the large trusses, the top of the dome will move twice the amount of the change of the cambre. The upper end should therefore be west of a vertical through the centre. When the wind is directly in the west, we have supposed that the pressure at the angle of the trusses due to this cause cannot exceed 15,000 lbs. on each, and hence the deflection in this case due to all the causes will be less than 1.704 inches; and the top of the dome will be moved to the east, somewhat less than $(2 \times 1.704 =)$ 3.408 inches. The investigation shows that the top of the dome should be made about $2\frac{1}{2}$ inches west of the centre of the base of the dome.

172.—IF THE BAYS IN THE CHORDS ARE EQUAL, as shown in Figs. 114 and 115, the strains upon the several parts may be expressed by simple formulas. In Fig. 114 the vertical members of the secondary trussing are ties, and the inclined pieces are braces; but in Fig. 115 the reverse is true, that is, the vertical members are struts and the inclined ones are ties. If the truss is composed entirely of iron, the form shown in Fig. 115 is preferable, because long pieces which are subjected to compression require proportionably more material than when subjected to tension (see Articles 21 and 22); and the inclined pieces in Fig. 114 are longer than the vertical ones in Fig. 115. But if the struts are made of wood and the ties of iron, the form shown in Fig. 114 will be more economical than that shown in Fig. 115.

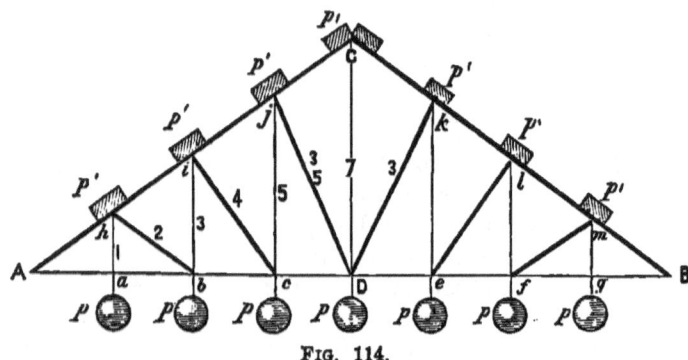

Fig. 114.

ANALYSIS OF FIG. 114.—A slight inspection of the problem shows:—

1st, That the strains upon the lower tie will be the same whether the load be upon the joints of the long tie, or upon the joints of the rafter, or upon both.

2d, The strains upon the braces of the secondary trussing, or upon the segments of the long rafters, will be the same whether the load be at the upper or lower joints.

3d, The strains upon the ties will be less when the load is upon the rafters than when it is upon the lower tie; for when upon the lower tie, the strains due to the weights are transmitted directly through the ties to the upper joints; but if the weights are at the upper joints they are supported directly by two unequally inclined rafters.

At first neglect the force of the wind, and suppose that equal weights are placed at each of the joints (or nodes) in the rafter and long tie.

Let N = the number of bays in the long tie AB;
p = the load at each of the joints of the tie;
p' = the load at each of the joints of the rafters;
n = the number of the bay considered;
V = the reaction at A;

SECONDARY TRUSSING.

c_n = the compression on the n-*th* division of the rafter; and

t_n = the tension upon the n-*th* bay of the tie.

Stress on the main rafters.—Take any point, as D, in the tie as the origin of moments; and from this point let fall a perpendicular on to the rafter AC. Suppose that a point x (not shown in the Fig.) is the foot of this perpendicular. If jC be severed, the system will turn about D, and the moment of stress on jC will be $c_n \times Dx$.

We also have

$V = \frac{1}{2}(N-1)(p+p')$;

$V \times AD$ = the moment of V in reference to D;

$(n-1)(p+p')$ = the load between A and D;

$\frac{1}{2} nl$ = lever arm of the preceding load;

$\frac{1}{2} n(n-1) l (p+p')$ = the moment of the load.

$\therefore c_n \times Dx = V nl - (n-1)(p+p') \frac{1}{2} nl$

$\therefore c_n = \frac{1}{2}[N-n](p+p') \frac{nl}{Dx}$

But the imaginary triangle ADx is similar to ADC, and hence

$$\frac{nl}{Dx} = \frac{AC}{CD}$$

$$\therefore c_n = \frac{1}{2}[N-n](p+p') \frac{AC}{CD} \dots\dots\dots(187)$$

In any practical case $\frac{AC}{CD}(p+p')$ will be constant. Call it q. Then

$$c_n = \frac{1}{2}[N-n] q = \frac{1}{2} Nq - \frac{1}{2} nq \dots\dots(188)$$

This is a maximum for $n = 1$, and decreases as n increases, hence the greatest stress on the main rafters is near the ends and least at the apex.

Stress on the long tie, AB. Call cD the n-th bay, and take j as the origin of moments, and we have

$V \times Ac = V \times (n-1) l$ = the moment of V;

$(n-1)(p+p')$ = the load from A to and including jc;

$\frac{1}{2}(n-2) l$ = the lever arm of the preceding load;

$\tfrac{1}{2}(n-1)(n-2)(p+p')l = $ the moment of the preceding load; and

$t_n \times j_i = $ the moment of the tension.

$$\therefore t_n = \tfrac{1}{2}\left[(N-1)-(n-2)\right]\frac{(n-1)l}{jc}(p+p')$$

But $\dfrac{(n-1)l}{jc} = \dfrac{AD}{DC}$

$$\therefore t_n = \tfrac{1}{2}\left[N-n+1\right]\frac{AD}{DC}(p+p')\ldots(189)$$

As before, let $\dfrac{AD}{DC}(p+p') = r$

$$\therefore t_n = \tfrac{1}{2}[N-n+1]r\ldots\ldots\ldots\ldots(190)$$

In this equation n must not be less than 2, for when $n = 2$ we find the stress on ab which is the same as that on Aa. The stress decreases as n increases.

Stress on the Vertical Ties.—The stress on the tie marked 1 is evidently p. The tie marked 3 virtually sustains one end of the king-post truss Ahb and the weight at b, and hence the stress on it is $\tfrac{1}{2}(p+p') + p = \tfrac{3}{2}p + \tfrac{1}{2}p'$. In the same way we find that the stress on the tie marked 5 (which is really the third tie) is two-thirds of the stress on 3, *plus* two-thirds of the weight at i, *plus* the weight at c; or equal to $2p + p'$. Similarly, we find that the stress on the n-th vertical tie is

$$\tfrac{1}{2}(n+1)p + \tfrac{1}{2}(n-1)p' =$$
$$\tfrac{1}{2}n(p+p') + \tfrac{1}{2}(p-p')\ldots\ldots\ldots\ldots(191)$$

from which it appears that the stress increases as n increases.

Stress upon the Braces.—The stress on the first brace (marked 2) is $\tfrac{1}{2}(p+p')\sec\theta$, in which θ is the inclination from the vertical. On the second it is $[\tfrac{1}{4}(p+p') + \tfrac{3}{4}(p+p')]\sec\theta = (p+p')\sec\theta$. On the third it is one-fourth of the load at a and h, *plus* two-fourths of the load at b and i, *plus* three-fourths of the load at c and j multiplied by $\sec\theta$; or $\tfrac{3}{2}(p+p')\sec\theta$. And generally, *the stress on* the n-th brace is

$$F_n = \tfrac{1}{2}n(p+p')\sec\theta\ldots\ldots\ldots\ldots\ldots(192)$$

in which θ is the inclination of a brace from the vertical. It

SECONDARY TRUSSING.

will be seen that the stress upon the braces increases directly as their distance from the support.

SECOND SOLUTION.—Conceive that a vertical section is made through the truss just at the right of j, and it will intersect the n-th division of the rafter, and of the tie and n-th brace (although the number of the n-th brace is one less than the number of the tie; thus, if the number of bay is 4, the number of the brace directly over it is number 3). It is then evident that the sum of the horizontal components of the strains in Dj and jC will equal the tension on the main tie.

Hence, $t_n = c_n \cos CAD + F \sin cjD$.

Also, the vertical shearing stress in the section equals the sum of all the stresses between A and j; or

$$V - (n - 1)(p + p') = c_n \sin CAD - F \cos cjD.$$

Also the moment of the tension taken about j as an origin, equals the sum of the moments of the applied forces; or

$$t_n \times jc = V.Ac - (p + p') ac - (p + p') bc - (p + p') C$$

$$\therefore t_n = \tfrac{1}{2}\left[N - n + 1 \right] \frac{AD}{DC}(p + p')$$

as before found.

We also have $\cos CAD = \dfrac{AD}{AC}$

$$\sin CAD = \frac{CD}{AC}$$

$$\cos cjD = \frac{jc}{jD}$$

$$\sin cjD = \frac{cD}{jD}$$

$$jc = \frac{DC}{AD} \cdot Ac$$

These values substituted in the preceding equations give, after eliminating F,

$$nc_n = \tfrac{1}{2}\left[(N - n) n \right] \frac{AC}{DC}(p + p')$$

$$\therefore c_n = \tfrac{1}{2}\left[N - n\right](p + p')\frac{AC}{DC},$$

As before given in Equation (187).

By eliminating c_n from the preceding equations, the value of F may be found.

ANALYSIS OF FIG. 115.—In this case, as has been before stated, the verticals are struts and the diagonals are ties.

Tension on the tie AB.—Suppose that bc (the n-th bay) is

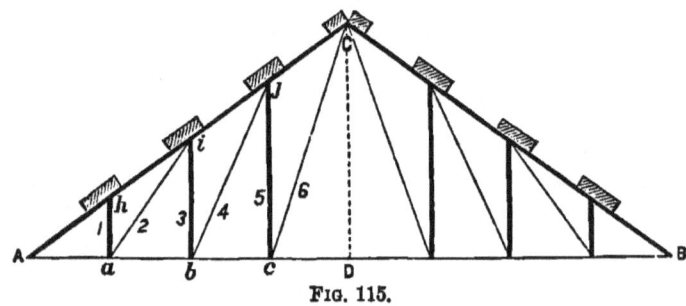

Fig. 115.

severed. The truss will then fail by turning about the joint j. Take j as the origin of moments, and we have

$$V \cdot nl - \tfrac{1}{2} n (p + p')(n - 1) l = t_n j c.$$

$$\therefore t_n = \tfrac{1}{2}\left[(N - 1) - (n - 1)\right]\frac{nl}{jc}(p + p')$$

$$= \tfrac{1}{2}\left[N - n\right]\frac{AD}{DC}(p + p')$$

Similarly, *the stress on the n-th bay of the rafter will be*

$$c_n = \tfrac{1}{2}\left[N - n + 1\right]\frac{AC}{DC}(p + p'),$$

The stress upon the n-th vertical strut will be,

$$\tfrac{1}{2}(n + 1) p' + \tfrac{1}{2}(n - 1) p$$

The stress upon n-th tie will be

$$\tfrac{1}{2} n (p + p') \sec \theta$$

in which θ is the inclination of a tie from the vertical.

SECONDARY TRUSSING.

EFFECT OF THE WIND IN THE PRECEDING CASE.—We must assume the direction of action of the wind when it strikes the roof. If the pressure is vertical we may include its pressure in p' in the preceding equations. If it be horizontal it will act only on one side, and tend to move the truss horizontally on its supports, and similarly for any other angle of action except a vertical one.

Let $\theta =$ the angle which the direction of the wind makes with the horizontal;

$i =$ the angle of the roof with the horizontal; and

$w =$ the pressure of the wind on each joint.

Then

$\theta - i =$ the angle of the direction of the wind with the roof; and

$w \sin (\theta - i) =$ the pressure perpendicular to the roof;

$w \cos (\theta - i) =$ the pressure parallel to the slope of the roof.

The latter value will give a pressure downward along the rafter when $\theta - i$ is less than 90°, and the reverse when $\theta - i$ exceeds 90°.

The perpendicular pressure may again be resolved into a vertical pressure (and may be represented by p'), and a longitudinal pressure which will produce compression or tension, the same in kind as that above stated.

The vertical pressure will be

$$w \sin (\theta - i) \div \cos i = p'$$

and the corresponding longitudinal component will be

$$w \sin (\theta - i) \times \tan g\, i$$

and hence the total longitudinal pressure will be

$$w \left[\cos (\theta - i) - \frac{\sin (\theta - i)}{\sin i} \right]$$

The vertical pressures must be treated as in the preceding cases, excepting that in determining the value of V we must observe that the pressure will generally be upon one side. It is not deemed necessary to give a complete analysis of this case, for all the principles which are necessary for its solution have been given in the preceding pages.

EXAMPLE. Suppose that the span is 80 feet; depth 16 feet; the load on the lower chord 24,000 lbs. uniformly distributed, and on the rafter 48,000 lbs. Required the stress upon the several parts of a truss like Fig. 114, when there are eight bays in the long tie.

$$\text{We have } p' = 2p = 6,000 \text{ lbs.}$$
$$p = 3,000 \text{ lbs.}$$
$$N = 8$$
$$AC = 43.08 \text{ feet.}$$
$$\therefore Aa = ab = bc, \text{ etc.} = 10 \text{ feet.}$$
$$ah = 4 \text{ feet.}$$
$$ib = 8 \text{ feet.}$$
$$jc = 12 \text{ feet.}$$
$$CD = 16 \text{ feet.}$$
$$hb = \sqrt{10^2 + 4^2} = 10.72.$$
$$ic = \sqrt{10^2 + 8^2} = 12.80.$$
$$jD = \sqrt{10^2 + 12^2} = 15.62.$$
$$\sec ahb = 2.68.$$
$$\sec bic = 1.60.$$
$$\sec cjD = 1.30.$$
$$\sec ACD = 2.69.$$

From Eq. (187) we have, by making $n = 1, 2, 3$, etc.
  stress on $Ah = \frac{1}{2} \times 7 \times 9,000 \times 2.69 = 84,735$ lbs.
  stress on $hi = 72,630$ lbs.
  stress on $ij = 60,525$ lbs.
  stress on $jC = 48,420$ lbs.

From Eq. (189) we have
  stress on $ab = 78,750$ lbs. = stress on Aa.
  stress on $bc = 67,500$ lbs.
  stress on $cD = 56,250$ lbs.

From Eq. (191) we have
  stress on $ah = 3,000$ lbs.
  stress on $bi = 7,500$ lbs.
  stress on $cj = 12,000$ lbs.
  stress on $CD = 16,500$ lbs.

which is the stress due to the load from A to D; and hence the total stress on CD is double this amount.

From Eq. (192) we have
  stress on $hb = \frac{1}{2} \times 9,000 \times 2.68 = 12,060$ lbs.
  stress on $ic = 14,400$ lbs.
  stress on $jD = 17,550$ lbs.

173.—ANOTHER FORM OF ROOF TRUSSING is shown in Fig. 117, in which the main rafter AC is trussed by several secondary trusses.

SECONDARY TRUSSING.

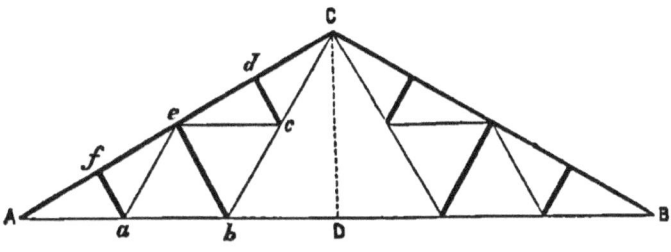

FIG. 117.

ABC is the **PRIMARY TRUSS**, in which AC and CB are the main rafters, and AB is the long tie rod. The rafter AC is supported at e by the inverted king-post truss AbC, which forms the first **SECONDARY TRUSS**, and in which eb is the strut and extends from the middle of AC and perpendicular to it, to where it intersects AB; and Ab and bC are the tie rods. Although Ab appears to be the same or a part of AB, yet in practice they are composed of separate rods.

The main rafter is still further trussed by the second set of secondary (or *tertiary*) trusses, Aae and ecC, in each of which the struts fa and dc are perpendicular to the main rafter and extend to the tie rods Ab and bC. In this construction it will be observed that ec is parallel to Ab, and ae parallel to bC.

In this truss it is supposed that all the load is on the main rafters, so that if the load is uniformly distributed over the rafters, we have

$W =$ the total load on the roof;
$N =$ the number of bays in both rafters; and
$\therefore \dfrac{W}{N} = p =$ the load at each of the joints $f, e, d,$ etc.

The load which is applied at d will be supported *direct'y* by the pieces Ad and dc, and according to Eqs. (88) and (89), Article 46, we will have for the

Stress on $ed = p \sin A$; and . . . . (193)

Stress on $dc = p \sin dce$;
$$= p \cos A = p \frac{AD}{AC} \dots \dots \dots \dots \dots (194)$$

The weight, p, at e produces *directly* the following stresses:—

Stress on $Ae = p \sin A$; and ....(195)

Stress on $eb = p \sin ebA$

$$= p \cos A = p \frac{AD}{AC} \dots \dots \dots (196)$$

But the strut eb also sustains one-half the pressures on dc and fa; and hence the total compression on the strut eb is

$$2 p \cos A \dots \dots \dots \dots (197)$$

For the load p at f we have

Stress on $fA = p \sin A$ ....(198)

Stress on $fa = p \sin faA$

$$= p \cos A = p \frac{AD}{AC} \dots \dots \dots (199)$$

For the load p at C, we have

Stress on $AC = \tfrac{1}{2} p \sec ACD$

$$= \tfrac{1}{2} p \frac{AC}{CD} \dots \dots \dots \dots (200)$$

But the stress on dc causes consequent stress on each of the parts ec, cC and eC, as shown from the inverted king-post; and hence we have for the stresses due to $p \cos A$ the following values:—

Stress on dC or $ed = \tfrac{1}{2} p \cos A \cot dCc = \tfrac{1}{2} p \dfrac{AD^2}{AC\cdot CD}$ (201)

Stress on Cc or $ec = \tfrac{1}{2} p \cos A \operatorname{cosec} dec = \tfrac{1}{2} p \dfrac{AD}{CD} \dots$ (202)

The expressions just found are also applicable to the secondary truss Aae.

For the secondary truss AbC we have in a similar manner for the load, or stress, $2 p \cos A$, which falls upon eb, the following values:—

Stress on Ae or $eC = p \dfrac{AD^2}{AC\cdot CD} \dots \dots \dots$ (203)

Stress on Ab or $bC = p \dfrac{AD}{CD} \dots \dots \dots$ (204)

SECONDARY TRUSSING. 221

Besides these stresses there is a stress the whole length of the rafters the same as if they were not trussed, which is due to the thrust at their upper ends. The amount of this compression is given by Eq. (53) and hence is (observing that W there equals $\frac{1}{2}$ W here)

$$\tfrac{1}{4} W \tang \theta \sin \theta, \text{ or}$$

$$= \tfrac{1}{4} W \frac{AD}{DC} \cdot \frac{AD}{AC} = \tfrac{1}{4} W \frac{AD^2}{DC.AC} \ldots \ldots (205)$$

This uniform load, as was shown in Article 20, causes a greater compression at the lower end of the several sub-rafters than at their upper end. Thus, the compression at the lower end of dC would be greater than at its upper end; but if the load be placed at the joints, as we are here considering, the compressions will be uniform from C to d, at which point it will receive an additional stress due to the load at d. It will then be uniform from d to e, where it will receive an additional stress due to the load at e, and so on. Although this view of the case is not strictly correct, yet in practical cases, where the parts Af, fe, etc., of the rafter are short, it is sufficiently exact.

In determining the total compression upon the rafter, we shall have somewhat too great a value if we add $\tfrac{1}{2} p \dfrac{AC}{DC}$ Eq. (200) to all the other strains found above, and somewhat too small a value if we omit it entirely; because the load being uniform there is no local load at the apex, and the load p placed at that point produces a greater strain than if it were uniformly distributed over the rafter. I have, however, retained it.

By collecting results, we have for the

Stress on $Af = \begin{cases} \text{the stress on } AC \text{ due to the thrust at } C, \\ + \text{ the stress on } AC \text{ due to the load } (p) \text{ at } C, \\ + \text{ the stress on } Ad \text{ due to the load at } d, \\ + \text{ the stress on } Ae \text{ due to the load at } e, \\ + \text{ the stress on } Ae \text{ due to the stress on } eb \text{ of the truss } AbC, \\ + \text{ the stress on } Af \text{ due to the load at } f, \\ + \text{ the stress on } Af \text{ due to the strain on } fa \text{ of the truss } Aae. \end{cases}$

The compression on *ed*, due to the truss *ecC*, is evidently not transmitted to *A*.

In a similar way we find the stress on any other part of the rafter. Hence we have

Total stress on $Cd = \frac{1}{4} W \frac{AD^2}{DC.AC} + \frac{1}{2} p \frac{AC}{CD} + \frac{3}{2} p \frac{AD^2}{AC.CD}$

Total stress

on $de = \frac{1}{4} W \frac{AD^2}{DC.AC} + \frac{1}{2} p \frac{AC}{CD} + \frac{3}{2} p \frac{AD^2}{AC.CD} + p \frac{CD}{AC}$,

on $ef = \frac{1}{4} W \frac{AD^2}{DC.AC} + \frac{1}{2} p \frac{AC}{CD} + \frac{3}{2} p \frac{AD^2}{AC.CD} + 2p \frac{CD}{AC}$

on $fA = \frac{1}{4} W \frac{AD^2}{DC.AC} + \frac{1}{2} p \frac{AC}{CD} + \frac{3}{2} p \frac{AD^2}{AC.CD} + 3p \frac{CD}{AC}$

on $fa = p \frac{AD}{AC} =$ stress on *dc*.

on $eb = 2p \frac{AD}{AC}$

on $Ab = p \frac{AD}{CD} =$ stress on *bC*.

on $Aa = \frac{1}{2} p \frac{AD}{CD} =$ stress on *ae*, *ec* and *cC*.

on $AB = \frac{1}{4} W \frac{AD}{DC}$

EXAMPLE. Let the span be 60 feet, and rise 15 feet. Also let the main rafters be supported at three points by secondary trussing, as in Fig. 117. Let the total load on the truss be forty tons. Required the stress on the several parts.

We have

$W = 40$ tons $= 80{,}000$ lbs.;
$N = 8$;
$p = 10{,}000$ lbs.;
$AD = 30$ feet;
$CD = 15$ feet;
$AC = 33.54$ feet;
$Ae = 16.77$ feet;
$Af = 8.385$ feet $= fe = ed = dC$;
$fa = 4.192$ feet;

for $AD : DC :: Af : fa \therefore fa = \frac{DC}{AD} \cdot Af = \frac{1}{2} Af$.

SECONDARY TRUSSING.

Similarly, $eb = 8.385$;
$Aa = 9.374 = ae = ec = Cc$; and
$Ab = 18.748 = bc.$

The stress on the long tie rod AB is

$$40,000 \times \tfrac{18}{9} = 80,000 \text{ lbs.}$$

The stress on Ab is, according to Eq. (204),

$$10,000 \times \tfrac{18}{9} = 20,000 \text{ lbs.}$$

The stress on Aa of the truss Aae is 10,000 lbs. = the stress on ae, ec, and eC.

The stress on $eb = 17,889$ lbs.
The stress on $Cd = 73,790$ lbs.
The stress on $de = 78,260$ lbs.
The stress on $ef = 82,730$ lbs.
The stress on $fA = 87,201$ lbs.

172.—THE PREVIOUS CASE MAY BE MODIFIED as shown in Fig. 118, in which the middle strut eb does not extend to the long tie rod. In this case the strains on the struts fa, eb, and dc, will be the same as in the preceding case, but the strains on the tie rods Ab, ae, etc., will be greater than in the preceding

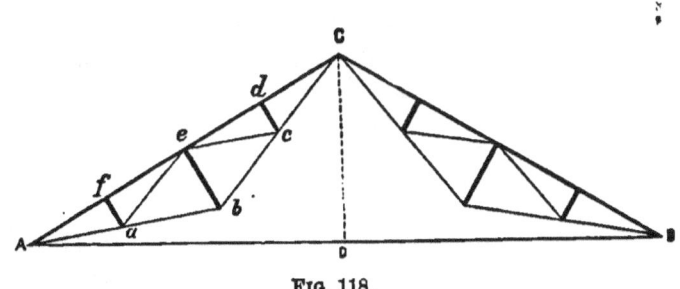

FIG. 118.

case, because they make a less angle with their base. For instance, the stress on eb ($2\, p \cos A$) will cause a stress of

$$p \sec e\, b\, A \cos A,$$

on the ties Ab and bC; and similarly for the others. The compression due to the total load on the rafter will be the same as in the preceding case; and the total strains will be found in the same manner as in the preceding case.

Part 4.

GENERAL PROBLEM OF TRUSSED GIRDERS.

We have thus far passed from the more simple cases of trussed girders to the more complex, determining, as we passed along, the effect due to each condition by itself. We may, however, proceed in the opposite way, and deduce all these results from the general equations of Statics.

175.—GENERAL EQUATIONS.—If all the forces which act upon a rigid body are resolved in the direction of three co-ordinate axes, we know that the sum of the forces will be zero, and the sum of the moments (or statical couples) will also be zero when they are in equilibrium among themselves. The forces which thus act upon a body are called *external forces*. If now we conceive that the several points which are thus acted upon are connected by rigid right lines, the same condition not only holds good, but we may find the strains (whether of tension or compression) upon these rigid right lines, by conceiving that one or all of them are severed, and instead of the tensions which transmit the stresses, we substitute forces which will produce the same effect upon the rigid lines. The forces which we thus substitute we call by way of distinction, *internal forces*. The internal forces may be treated in all respects like external forces.

Suppose that a frame of any kind whatever, as Fig. 119, is so made as to connect all the points of application of the external forces; and that a plane section is made so as to cut several of the bars.

Let P, P_1, P_2, P_3, etc., be the external forces;
F, F_1, F_2, F_3, etc., be the internal forces.

TRUSSED GIRDERS.

Take any convenient point for the origin of rectangular co-ordinates, and let x and z be horizontal, and y vertical, as in Fig. 119.

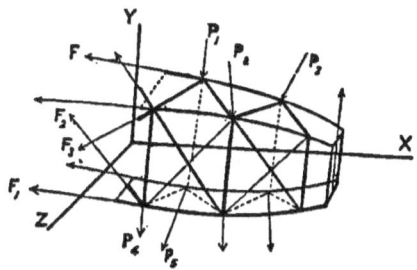

FIG. 119.

Let x, x_1, x_2, etc.
y, y_1, y_2, etc.   be the co-ordinates of the point of application of the forces;
z, z_1, z_2, etc.

a, a_1, a_2, etc., be the angles which the external forces make with the axis of x;

β, β_1, β_2, etc., be the angles which they make with axis of y; and

$\gamma, \gamma_1, \gamma_2$, etc., be the angles which they make with the axis of z,

a, a_1, a_2, etc.
b, b_1, b_2, etc.   be the corresponding angles made by the internal forces with the axes x, y, and z.
c, c_1, c_2, etc.

Then for equilibrium, we have

$$\left.\begin{array}{l} P\cos a + P_1\cos a_1 + P_2\cos a_2 + \text{etc.}, + F\cos a + F_1\cos a_1 + \text{etc.} = 0 \\ P\cos \beta + P_1\cos \beta_1 + P_2\cos \beta_2 + \text{etc.}, + F\cos b + F_1\cos b_1 + \text{etc.} = 0 \\ P\cos \gamma + P_1\cos \gamma_1 + P_2\cos \gamma_2 + \text{etc.}, + F\cos c + F_1\cos c_1 + \text{etc.} = 0 \end{array}\right\} (a)$$

or, more briefly,

$$\left.\begin{array}{l} \Sigma P \cos a + \Sigma F \cos a = 0 \\ \Sigma P \cos \beta + \Sigma F \cos b = 0 \\ \Sigma P \cos \gamma + \Sigma F \cos c = 0 \end{array}\right\} \dots\dots\dots\dots\dots\dots(b)$$

We also have

$$\left.\begin{array}{l} \Sigma P (x \cos \beta - y \cos a) + \Sigma F (x \cos b - y \cos a) = 0 \\ \Sigma P (z \cos a - x \cos \gamma) + \Sigma F (z \cos a - x \cos c) = 0 \\ \Sigma P (y \cos \gamma - z \cos \beta) + \Sigma F (y \cos c - z \cos b) = 0 \end{array}\right\} \dots\dots(c)$$

We thus have six independent equations of the most general form, and by means of which we may determine six unknown quantities. If, therefore, in the most general form of truss, the plane section cuts more than six bars of the truss, the solution in regard to the strains is indeterminate, unless conditions can be established among them which will give other equations. For instance, certain bars may receive equal strains, certain others may be strained half as much, or some ratio may be established among them.

But few practical problems of such a general character ever occur, but it is easy to conceive of theoretical ones like the following.

EXAMPLE.—A truss is being raised into position, and is held by ropes, which are inclined in various directions, while the truss is acted upon by its own weight, the force of the wind and the ropes. It is required to find the condition of equilibrium among the external forces; and the stress upon all the bars in a plane section.

176.—FORCES IN A PLANE.—In nearly all engineering structures we have only to consider forces in a plane, since the force of gravity is the chief force with which we have to contend.

Suppose that the forces are all in the plane xy. Then all the components and moments in regard to z reduce to zero, and the preceding Equations become

$$\left. \begin{array}{l} \Sigma\, P \cos a + \Sigma\, F \cos a = 0 \\ \Sigma\, P \cos \beta + \Sigma\, F \cos b = 0 \\ \Sigma\, P\, (x \cos \beta - y \cos a) + \Sigma\, F\, (x \cos b - y \cos a) = 0 \end{array} \right\} (d)$$

This gives us three independent Equations, and hence the problem is determinate when the vertical section cuts only three bars.

These Equations may be developed so as to be especially applicable to particular cases.

177.—APPLIED FORCES ALL VERTICAL.—In this case a will be 90° or 270°; and $\beta = 0°$ or 180°

$$\therefore \Sigma\, P \cos \beta = \Sigma \pm P$$

If $\Sigma\, V =$ the forces which act vertically upward, and $\Sigma\, P =$ those which act vertically downward, we have

$$\Sigma\, P \cos \beta = \Sigma\, P - \Sigma\, V$$

In most mechanical structures the forces which act vertically upward are the resisting forces at the supports.

We also have for this case
$$\Sigma\, P \cos a = 0$$
And hence Equations (d) become
$$\left.\begin{array}{l}\Sigma\, F \cos a = 0 \\ \Sigma\, P - \Sigma\, V + \Sigma\, F \cos b = 0 \\ \Sigma\, P.x - \Sigma\, V.x + \Sigma\, F(x \cos b - y \cos a) = 0\end{array}\right\}\dots\dots(e)$$

The quantity ($x \cos b - y \cos a$) is the lever arm of the force F.

Let q, q_1, etc., = the lever arms of the forces;
F = the stress on the upper bar in the section;
F_1 = the stress on the lower bar in the section; and
F_2 = the stress on the intermediate bar.

Then Equations (e) become
$$\left.\begin{array}{l}F \cos a + F_1 \cos a_1 + F_2 \cos a_2 = 0 \\ \Sigma P - \Sigma V + F \sin a + F_1 \sin a_1 + F_2 \sin a_2 = 0 \\ \Sigma P.x - \Sigma V.x + F.q + F_1 q_1 + F_2 q_2 = 0\end{array}\right\}\dots(f)$$

If the origin of moments is taken at the intersection of upper chord and the intermediate bar, q and q_2 become zero; and the third of the preceding Equations becomes
$$\Sigma\, P.x - \Sigma\, V.x + F_1 q_1 = 0.$$

EXAMPLES. 1. Required the inclination of the chords and of the intermediate piece, so that the stress shall be the same on all of them at all sections for a uniform load over the whole length.

We have $F = F_1 = F_2$ and the Equations become

$$\begin{array}{l}\cos a + \cos a_1 + \cos a_2 = 0 \\ \Sigma P - \Sigma V + F \sin a + F \sin a_1 + F \sin a_2 = 0 \\ \Sigma P x - \Sigma V.x \quad\quad + F q_1 \quad\quad\quad = 0\end{array}$$

which Equations have three unknown angles and one unknown stress; and hence the solution is indeterminate.

2. Let the data be as in the preceding problem, and the lower chord horizontal, or $a_1 = 0$ degrees.

$$\begin{array}{l}\cos a + 1 + \cos a_2 = 0 \\ \therefore\, \Sigma P - \Sigma V + F \sin a + 0 + F \sin a_2 = 0 \\ \Sigma P.x - \Sigma V.x \quad\quad + F q_1 \quad\quad = 0\end{array}$$

From the second and third of these Equations we have
$$F = \frac{-\Sigma Px + \Sigma Vx}{q_1} = \frac{-\Sigma P + \Sigma V}{\sin a + \sin a_2}$$

which combined with the first of the preceding set of equations enables one to find a and a_2 when q_1 is known. But we cannot assume arbitrary values for q_1 (the depth of the truss) without affecting the value of a. There are therefore really four unknown quantities, and it is necessary to establish another Equation depending upon the form of the truss to make the solution determinate.

The load being uniform,

Let $w =$ the load per foot of length;
$L =$ the span; and
$x =$ any variable distance from the support;

Then $\Sigma\ V = \tfrac{1}{2} wL$;
$\Sigma\ P = wx$;
$\Sigma\ Vx = \tfrac{1}{2} wLx$; and
$\Sigma\ Px = \tfrac{1}{2} wx^2$

Hence the preceding Equation becomes

$$\sin a + \sin a_2 = \frac{-2x + L}{Lx - x^2} q_1$$

At the middle, $x = \tfrac{1}{2} L$ and we have

$$\sin a + \sin a_2 = 0$$

or, $a = -a_2$ or $180° + a_2$

3. The lower chord being horizontal and uniformly loaded over its whole length, it is required to find the inclination of the upper chord and braces so that that they shall be equally strained.*

Take the origin of co-ordinates at the point where the intermediate piece intersects the upper chord, and we have $q = q_2 = 0$.

The load being uniform, we have (w, L and x being the same as in the preceding problem)

$$\Sigma\ V = \tfrac{1}{2} wL$$
$$\Sigma\ P = wx$$
$$\Sigma\ Vx = \tfrac{1}{2} wLx$$
$$\Sigma\ Px = \tfrac{1}{2} wx^2$$

Also let

$H_1 =$ the stress on the lower chord; and
$h =$ the depth of the truss at the section considered.

We also have

$$F = F_2,\ F_1 = H,\ \text{and } \cos a_1 = 1$$

Hence Equations (f) become

$$F \cos a + H_1 + F \cos a_2 = 0$$
$$wx - \tfrac{1}{2} wL + F \sin a + 0 + F \sin a_2 = 0$$
$$\tfrac{1}{2} wx^2 - \tfrac{1}{2} wLx \quad\quad + H_1 h \quad\quad = 0$$

---

* Solution by G. W. Mickle, class of 1870, *Univ. of Mich.*

TRUSSED GIRDERS. 229

By equating the values of F derived from the first and second of these Equations, we have

$$\frac{H_1}{\cos a + \cos a_2} = \frac{\frac{1}{2} w (L - 2x)}{\sin a + \sin a_2} \dots \dots \dots \dots \dots (a)$$

The third Equation above gives

$$H_1 = \frac{\frac{1}{2} wx (L - x)}{h} \dots \dots \dots \dots \dots \dots \dots \dots \dots (b)$$

In these Equations are three unknown quantities, h, a, and a_2, and hence the solution is indeterminate unless another relation is established between them. This may be done by assuming a depth for the truss at the middle, and dividing the lower chord into a number of equal bays.

Let $D =$ the depth of the truss at the middle;
$2 N =$ the number of bays in the lower chord;
$L =$ the length of the span;
$l = L \div 2 N =$ the length of one bay;
$a_n =$ the angle which the n-th bay of the upper chord makes with the horizontal; and
$n =$ the number of the bay considered.

Taking the origin of co-ordinates at the middle of the upper chord, we have

$$h = D - l \left[\tang a_{n-1} + \tang a_{n-2} + \text{etc.} \dots \tang a_{n-s} \right]$$

$$\tang a_n = \frac{l}{h} = \frac{l}{D - l \left[\tang a_{n-1} + \tang a_{n-2} \text{ etc.} \right]}$$

$$x = (N - n + 1) l.$$

Eq. (b) becomes

$$H_1 = \frac{\frac{1}{2} w (N - n + 1) l (L - (N - n + 1) l)}{D - l \left[\tang a_{n-1} + \tang a_{n-2} + \text{etc.} \dots \tang a_{n-s} \right]}$$

After substituting these values in Eq. (a), it may be reduced to the form

$$m \cos a + r \sin a = Q$$

in which m, r and Q are known.

This may be solved by the introduction of an auxiliary angle ϕ (see "Chauvenet's Trigonometry," p. 90) by putting

$$k \sin \phi = m; \text{ and}$$
$$k \cos \phi = r$$
$$\therefore \sin (\phi + a) = \frac{Q}{k} = \frac{Q}{m} \sin \phi = \frac{Q}{r} \cos \phi$$

As a check upon the calculation, we have

$$\tang a = \frac{h_n - h_{n-1}}{l}$$

NUMERICAL EXAMPLE. Let $N = 4$; $L = 240$ $D = 10$, and $w =$ half ton.

We have

For $n = 1$, $x = 120$, $h = 10$. $H_1 = 360.00$, $\theta = 108° 26'$ $a = 341° 26'$
   $n = 2$, $x = 90$, $h = 19.999$, $H_1 = 168.77$, $\theta = 51° 18'$ $a = 313° 59'$
   $n = 3$, $x = 60$, $h = 51.08$, $H_1 = 52.85$, $\theta = 30° 25'$ $a = 329° 33'$
   $n = 4$, $x = 30$, $h = 68.72$, $H_1 = 38.2$, $\theta = 23° 35'$.

A construction of these results shows a peculiar form of truss.

4. Required the inclination of the upper chord for a uniform load so that the stress on the upper chord shall be uniform throughout, for the panel system, as shown in Fig. 105.

The equal bays in the lower chord being known, and any depth as $G O = h$ Fig. 105, being assumed, we may find a_2 which substituted in Eq. (f) (making q and $q_2 = o$ and $a_2 = o$) gives three Equations, from which we may find F_1 a and F_2 in terms of F.

Having found a we may readily find HR, and proceed as before.
The value of F may be found at the middle by an equation of moments.

Resuming Eqs. (f), we proceed to fix more definitely the values of some of the quantities in them.

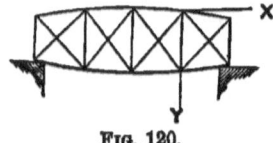

FIG. 120.

If the origin of co-ordinates be taken at the intersection of the upper chord and intermediate bar, we have found, page 227, that

$$F_1 q_1 = \Sigma V.x - \Sigma P.x$$

If $h = $ the depth of the truss in the section which is considered, and a_1 the inclination of the lower chord to the axis of x, as has thus far been assumed, we have the perpendicular from the *origin of moments*[*] to the lower chord.

$$q_1 = h \cos a_1$$

It is not generally necessary to consider more than one force at the support, which call V.

If the origin of co-ordinates be at the middle of the upper chord, and the span be L, the arm of V will be $\frac{1}{2} L - x$, and the particular value of $\Sigma P.x$ can be determined when the conditions of the loading are completely known.

---

[*] The origin of *moments* may be at any distance from the origin of co-ordinates.

The first of Eqs. (f) gives

$$F \cos a + F_2 \cos a_2 = - F_1 \cos a_1$$

hence, *the algebraic sum of the horizontal components of the stresses in the upper chord and the intermediate piece equals the horizontal component of the stress in the lower chord taken with a contrary sign;* or, in other words, *the resultant horizontal compression at the joint in the upper chord equals the resultant horizontal tension in the lower chord.*

If we call tension *plus* and compression *minus*, we shall generally have $F \cos a$ positive, and $F_2 \cos a_2$ negative; and $F_1 \cos a_1$ will be positive or negative, according as it is a tie or brace. If, however, we proceed as is common in mechanics, by considering all forces as positive and attributing the proper values to the signs, it will not be necessary to consider whether a strain is compressive or tensive, but only the *direction* in which it must act to produce the strain.

The second of Eqs. (f) may be reduced to

$$F \sin a + F_1 \sin a_1 + F_2 \sin a_2 = \Sigma V - \Sigma P$$

the second member of which is called the VERTICAL SHEARING STRESS; hence the vertical shearing stress equals the sum of the vertical components of all the stresses in the section considered.

By comparing Figs. 120 and 106, we see that if a_1 in Fig. 120 is positive, and less than 90°; that in Fig. 106 for the corresponding part of the truss will be negative and less than 90°; or it may be considered positive and between 270 and 360°. These conditions do not change the sign of $\cos a_1$ but $\sin a_1$ will have contrary signs in the two cases.

178.—LOWER CHORD HORIZONTAL.—In this case $a_1 = 0°$ or 180°, and still considering the forces as vertical, and Eqs. (f) are directly applicable to this case. In order to make the notation conform with that previously used, let

$a = i =$ the angle which the upper chord makes with the horizontal.

$90° - a_2 = \theta =$ the angle which the intermediate piece makes with the vertical; and

$t_n =$ the stress in the horizontal lower chord;

$c_n =$ the stress in the upper chord;

And, taking the origin of co-ordinates at a joint in the upper chord, Eqs. (f) become

$$\left. \begin{array}{l} c_n \cos i + F_2 \sin \theta = t_n \\ c_n \sin i + F_2 \cos \theta = V - \Sigma P \\ t_n h = V x - \Sigma P . x \end{array} \right\} \quad (g)$$

If the intermediate piece is a *brace*, $\sin \theta$ will be positive; but if it be a *tie*, $\sin \theta$ will be negative, and the Equation will become

$$c_n \cos i - F_2 \sin \theta = t_n$$

which is the same as Eq. (177) page 182.

The second of Eqs. (g) is the same as Eq. (178) page 183, and the third of Eqs. (g) would give Eq. (176) when reduced for that case.

Eqs. (g) are not only applicable to the parabolic arched truss as developed on pages 181 to 186, inclusive, but to all cases in which the lower chord is horizontal. They are applicable to roof trusses in which the angle i is constant.

179.—UPPER CHORD HORIZONTAL.—For this case $i = 0$ or $180°$, according as x is positive to the right or left, and Eqs. (f) become

$$\left. \begin{array}{l} t_n \cos i_1 + F_2 \sin \theta = c_n \\ t_n \sin i_1 + F_2 \cos \theta = V - \Sigma P \\ t_n h \cos i_1 = V x - \Sigma P x \end{array} \right\} \quad (h)$$

The *amount* of the strains in this case is the same as for a truss in the preceding case inverted.

180.—BOTH CHORDS HORIZONTAL.—For this case $i = 0$ and $i_1 = 180°$, or the reverse, and Eqs. (g) or (h) become

$$\left. \begin{array}{l} c_n + F_2 \sin \theta = t_n \\ F_2 \cos \theta = V - \Sigma P \\ t_n h = V x - \Sigma P . x \end{array} \right\} \quad \ldots\ldots\ldots\ldots\ldots (i)$$

These Eqs. cover all cases of parallel chords whether of the *panel* or *triangular* systems. There being three equations, there may be three unknown quantities, and the solution be determinate; but if there are more than three unknown quantities, other relations must be established.

From the first of Eqs. (i) we have

$$t_n - c_n = F_2 \sin \theta, \ldots\ldots\ldots\ldots\ldots (j)$$

Hence the difference in the stresses in the upper and lower chords in the same vertical, equals the horizontal component of the stress in the intermediate piece (brace or tie).

The second shows that the vertical component of the stress in the brace equals the *vertical shearing* stress; and if θ and V are constant, F_s will be greatest at the ends, and diminish as we pass from the end to the point where the shearing stress is zero. This result introduced into Eq. (j) shows that the strains in the upper and lower chords approach an equality as we pass from the end to the point where the vertical shearing stress is zero, at which point the strains will be the same in both chords.

The depth being constant, the third of Eqs. (i) shows that the stress in the lower chord varies directly as the resultant moment of the external forces.

Observing that the P's in the second number are constant, and differentiating, we have

$$\frac{d\,(t_n\,h)}{dx} = V - \Sigma\,P.$$

the second member of which is the same as the second member of the second of Eqs. (i), hence *the first differential coefficient of the moments of the applied forces, equals the vertical shearing stress, both taken in the same plane.*

If we consider the panel system,

and let $N =$ the number of bays in the lower chord;
$p =$ the load at each joint;
$n =$ the number of the bay considered, counting from the end;
$l =$ the length of a bay; and

we have $V = \frac{1}{2}\,(N - 1)\,p$

$\Sigma\,P = (n - 1)\,p$

$V.x = \frac{1}{2}\,(N - 1)\,pnl$

$\Sigma\,Px = (n - 1)\,p \times \frac{1}{2}\,nl = \frac{1}{2}\,(n - 1)\,npl$

Hence, Eqs. (i) become

$c_n + F_s \sin\theta = t_n$

$$F_s \cos \theta = \tfrac{1}{2} (N - 2n + 1) p$$
$$t_n h = \tfrac{1}{2} npl (N - n)$$
$$\therefore t_n = \frac{n(N-n)pl}{2h}$$

which is essentially the same as Eq. (136); and

$$F_s = \tfrac{1}{2} (N - 2n + 1) p \sec \theta$$

which is the value of the second term of Eq. (128).

181.—CASE OF A HORIZONTAL BEAM under the action of forces which are perpendicular to its axis.

Returning to Eqs. (e), let F be the resultant of the internal forces at any point. Take the origin of co-ordinates at the centre of the section, and let x coincide with the axis of the beam (or, as before, be perpendicular to the direction of the applied forces, the forces being perpendicular to the beam).

Then the first of Eqs. (e) shows that the algebraic sum of all the forces which act along the beam is zero, or, in other words, the sum of the compressions equals the sum of the tensions. This principle enables us to determine the position of the neutral axis. (See "Resistance of Materials.")

The second of Eqs. (e) shows that the sum of the vertical forces in a beam equals the resultant vertical applied forces between the section and the end; and the third shows that the sum of the moments of the internal forces equals the sum of the moments of the external forces.

In order to determine the moments of F it is necessary to know the *law of action* of the internal forces.

One of the principal laws is:—the strains vary directly as their distance from the neutral axis, and gives rise to the expression

$$R \frac{I}{d_1}$$

in which R = the modulus of resistance to transverse strains;
I = the moment of inertia of the transverse section; and
d_1 = the distance from the neutral axis to the most remote fibre.

Hence, the third Equation becomes

$$R\frac{I}{d_1} = \Sigma\, Vx - \Sigma\, Px$$

Having found R, the value of $\Sigma\, F \cos a$ may be found from the law just given. These cases are fully discussed in the Author's "Resistance of Materials."

If the forces are applied in the direction of the length of the piece, we will have in Eqs. (d)

$$a = 0;\ a = 0$$
$$\beta = 90°;\ b = 90°$$

and the Equations become

$$\Sigma\, P = \Sigma\, F$$
$$\Sigma\, Py = \Sigma\, Fy$$

which are applicable to a column under flexure, and other cases in which the applied and resisting forces are parallel, but not coincident. If they are coincident, $y = 0$, and we have only the first of the two Equations remaining, which is directly applicable to the elongation and compression of elastic pieces. (See "Resistance of Materials.")

182.—PERFECTLY FLEXIBLE SYSTEMS.—If a rope or other perfectly flexible continuous physical line be secured at two points, and loaded continuously between those points according to any law, the flexible string will assume some definite curvilinear form.

When the load is the weight of the string only, the curve is called a "catenary." Suppose that the string is fixed at its extremities, Fig. 121, and is acted by any system of continuous forces. Take the origin of co-ordinates at any point as C: x and z horizontal, and y vertical.

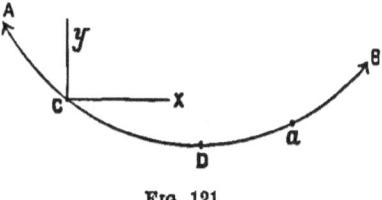

Fig. 121.

Let $t =$ the tension at any point, as a;
  $t_0 =$ the tension at the origin, C;
  $X_0 =$ the horizontal component of the tension at C;

Y_o = the vertical component of the tension at the same point;

Z_o = the horizontal component of the tension at the same point in a direction perpendicular to xy; and

X, Y and Z, the corresponding components of all the applied forces between C and a;

s = any portion of the arc, as $C\,a$.

Then $\dfrac{dx}{ds}$, $\dfrac{dy}{ds}$, $\dfrac{dz}{ds}$ will be the cosines of the angles which the curve makes with the respective co-ordinate axis; and

$t\,\dfrac{dx}{ds}$ = the component of the tension in the direction of the axis of x; and

$t\,\dfrac{dz}{ds}$, $t\,\dfrac{dy}{ds}$ similar values for the axis z and y; and

Eqs. (b) become

$$\left. \begin{array}{l} X + X_o + t\,\dfrac{dx}{ds} = 0 \\ Y + Y_o + t\,\dfrac{dy}{ds} = 0 \\ Z + Z_o + t\,\dfrac{dz}{ds} = 0 \end{array} \right\} \ldots\ldots\ldots(k)$$

It may be shown that Eqs. (c) give no new relations for this case, and hence Eqs. (k) are sufficient. These Eqs. are general, and would give the form of the curve of the string in all cases if the law of action of the forces were known, and the Equations could be integrated.

For instance, if a heavy string were placed in a stream of running water, the forces which act upon it would be gravity, or the weight of the string, which would be uniform along the length of the string (if the string were of uniform size), and the force of the running water, the law of action of which might not be known. If, however, we assume that the line joining the fixed points is the axis of x and is inclined at an angle of 45° with the direction of action of the water, and y is vertical, and that the weight of the string is w per unit of length, and that the pressure of the water is p on each unit of length of the projection of the string upon a plane which is perpendicular to the axis of the stream; then Eqs. (k) become

$$\left. \begin{array}{r} \cos 45° \, p \displaystyle\int_0^s \cos\left(45° + \cos^{-1}\dfrac{dx}{ds}\right) ds + X_o + t\dfrac{dx}{ds} = 0 \\[6pt] + ws + Y_o + t\dfrac{dy}{ds} = 0 \\[6pt] \cos 45° \, p \displaystyle\int_0^s \cos\left(45° + \cos^{-1}\dfrac{dz}{ds}\right) ds + Z_o + t\dfrac{dz}{ds} = 0 \end{array} \right\}$$

which are the differential Equations of the curve.

183.—FORCES IN A PLANE.—If the applied forces are all in a plane, let xy be that plane; then Eqs. (k) become

$$\left. \begin{array}{r} X + X_o + t\dfrac{dx}{ds} = 0 \\[6pt] Y + Y_o + t\dfrac{dy}{ds} = 0 \end{array} \right\} \quad \ldots\ldots\ldots\ldots (l)$$

If the applied forces are normal to the arc, the tension on the string will be uniform throughout; for there will be no tangential component among them to change the tension.

Let $p =$ the force applied at any unit of length of the arc, which may be constant, or vary as some function of the arc; and

$P =$ the total force applied to the arc.

Then generally

$$P = \int p \, ds;$$

$$\text{and } X = \int p \, ds \left(\dfrac{dx}{ds}\right); \text{ and}$$

$$Y = \int p\, ds \left(\frac{dy}{ds}\right)$$

Hence, Eqs. (*l*) become

$$\int p\, dx + X_o + t\frac{dx}{ds} = 0$$

$$\int p\, dy + Y_o + t\frac{dy}{ds} = 0$$

Differentiating these, gives

$$p\, dx + t\, d\left(\frac{dx}{ds}\right) = 0$$

$$p\, dy + t\, d\left(\frac{dy}{ds}\right) = 0$$

Squaring these, and adding, gives

$$p^2 = \frac{t^2}{ds^2}\left[\left(d\left(\frac{dx}{ds}\right)\right)^2 + \left(d\left(\frac{dy}{ds}\right)\right)^2\right] = \frac{t^2}{\rho^2}$$

in which ρ is the radius of curvature at the point whose co-ordinates are x and y.

This becomes

$$\rho = \frac{t}{p};$$

that is, *the radius of curvature of a normally pressed arc varies inversely as the pressure per unit of length at that point*. But as in such cases the pressure is not uniform over the whole length of the unit, we should say that it is what the pressure would be per unit of length if the pressure were the same that it is at the point considered.

If the arc is a circle, ρ is constant, and hence the pressure is constant.

If now the origin of co-ordinates be taken at the lowest point of the curve, or where it is horizontal, we shall have $Y_o = 0$, and Eqs. (l) become

$$X + X_o + t \frac{dx}{ds} = 0$$

$$Y + t \frac{dy}{ds} = 0$$

And if all the applied forces are vertical, X will be zero, and we have

$$\left. \begin{array}{r} X_o + t \dfrac{dx}{ds} = 0 \\ Y + t \dfrac{dy}{ds} = 0 \end{array} \right\} \quad \ldots\ldots\ldots (m)$$

From the first of Eqs. (m) we have

$$t \frac{dx}{ds} = X_o = t_o \; (t_o \text{ being the tension at the lowest point}),$$

that is, the horizontal component of the tension is constant and equal to that at the lowest point.

From the second we have

$$t \frac{dy}{ds} = -Y$$

that is, the vertical component of the tension equals the total load between the lowest point and the point where the tension is considered.

The form of the curve in these cases can readily be found when the law of the loading is known.

EXAMPLE 1. If the load be uniform over the horizontal, as is practically the case with the suspension bridge, we have

$$Y = wx$$

in which w is the load per unit of length; and Eqs. (m) become

$$t \frac{dx}{ds} = t_o$$

$$t \frac{dy}{ds} = wx$$

Dividing the latter by the former gives

$$\frac{dy}{dx} = \frac{wx}{t_o}$$

Integrating once gives

$$\frac{2 t_0}{w} y = x^2 + (C = 0)$$

which is the Equation of the common parabola. As t_0 enters into the parameter only, we see that the tension at the lowest point will always be the same, for the same parameter and load per unit of length. In this respect the tension at the lowest point is independent of the span.

We also have

$$t = t_0 \frac{ds}{dx} = t_0 \frac{\sqrt{dx^2 + dy^2}}{dx} = t_0 \sqrt{1 + \frac{dy^2}{dx^2}}$$

$$= t_0 \sqrt{1 + \frac{w^2 x^2}{t_0^2}}$$

which gives the tension at any point of the curve. The tension at the lowest point may be found by substituting the co-ordinates of the other extremity.

2. If the load is a continuous function of the length of the arc, we have

$$Y = w s$$

and Eqs. (m) become

$$t \frac{dx}{ds} = t_0$$

$$t \frac{dy}{ds} = w s$$

$$\therefore \frac{dy}{dx} = \frac{w}{t_0} s$$

Differentiating gives

$$d\left(\frac{dy}{dx}\right) = \frac{w}{t_0} ds = \frac{w}{t_0} \sqrt{dx^2 + dy^2}$$

$$= \frac{w}{t_0} dx \sqrt{1 + \frac{dy^2}{dx^2}}$$

$$\therefore \frac{w}{t_0} dx = \frac{d\left(\frac{dy}{dx}\right)}{\sqrt{1 + \frac{dy^2}{dx^2}}}$$

The first integral

$$\frac{w}{t_0} x = Nap. \log. \left[\frac{dy}{dx} + \sqrt{1 + \frac{dy^2}{dx^2}}\right]$$

$$\therefore e^{\frac{w}{t_0} x} = \frac{dy}{dx} + \sqrt{1 + \frac{dy^2}{dx^2}} = \frac{dy}{dx} + \frac{ds}{dx}$$

$$\text{or, } 1 + \frac{dy^2}{dx^2} = \left[e^{\frac{w}{t_0}x} - \frac{dy}{dx} \right]^2$$

which reduced gives

$$\frac{dy}{dx} = \tfrac{1}{2}\left[e^{\frac{w}{t_0}x} - e^{-\frac{w}{t_0}x} \right]$$

which integrated gives

$$y = \tfrac{1}{2}\frac{t_0}{w}\left[e^{\frac{w}{t_0}x} + e^{-\frac{w}{t_0}x} \right] + \left(C = -\frac{t_0}{w} \right)$$

$$= \tfrac{1}{2}\frac{t_0}{w}\left[e^{\frac{w}{2t_0}x} - e^{-\frac{w}{2t_0}x} \right]^2$$

184.—AN INVERSE PROBLEM.—If the form of the curve is known, we may determine the law of loading so that it shall be a curve of equilibrium; or, in other words, the resultant of all the forces at any point shall be in the direction of a tangent to the curve at that point.

Suppose that the loading is of uniform density, and that the increase or decrease, as the case may be, of the pressures is caused by variations in the depth of the loading.

Let d = the depth of the loading at the origin;
Z = the depth at any other point; and
δ = the weight per unit of volume of the loading.

Then Eqs. (m) become

$$t\,\frac{dx}{ds} = t_0$$

$$t\,\frac{dy}{ds} = \delta \int Z dx$$

Dividing the latter by the former gives

$$\frac{dy}{dx} = \delta\,\frac{\int Z dx}{t_0}$$

Differentiating gives

$$\frac{d^2y}{dx^2} = \frac{\delta Z}{t_0}$$

We know that

$$\rho = \frac{\left(1 + \frac{dy^2}{dx^2}\right)^{\frac{3}{2}}}{\frac{d^2y}{dx^2}} = \frac{sec^3 i}{\frac{d^2y}{dx^2}}$$

in which ρ is the radius of curvature, and $i =$ the angle which the curve makes with the axis of x.

Combining the preceding Equations gives

$$Z = \frac{t_o}{\delta} \frac{sec^3 i}{\rho}$$

At the origin we have $\rho = \rho_o$; $i = 0$; and $Z = d$.

$$\therefore d = \frac{t_o}{\delta \cdot \rho_o} \text{ or } \frac{t_o}{\delta} = d\rho_o$$

$$\therefore Z = d \, \rho_o \frac{sec^3 i}{\rho}$$

EXAMPLE. 1.—If the curve is the arc of a circle, $\rho = \rho_o$ and we have $Z = d \, sec^3 i$.

At the extremity of the quadrant we have

$$i = 90° \therefore Z = \infty$$

Hence it is practically impossible to make a full-centred arch a curve of equilibrium.

2. Let the curve be a parabola.
Then we have

$$x^2 = 2 \, p \, y,$$

for the Equation of a parabola, and

$$\rho = \frac{(x^2 + p^2)^{\frac{3}{2}}}{p^2}$$

$$\rho_o = p$$

and $sec^3 i = \left[\frac{x^2 + p^2}{p^2}\right]^{\frac{3}{2}}$

Hence, by substitution we find

$$Z = d,$$

or the depth of the loading must be constant, or uniformly distributed over the span, as was assumed in the first example of the preceding article.

THE END.

APPENDIX I.

DEMONSTRATIONS OF A CERTAIN GRAPHICAL SOLUTION.

Note referred to on Page 56. Suppose that several forces are applied at each of the angles of the triangle ABC, Fig. 122, and are in equilibrium among themselves. Let P_1, P_2 and P_3, be the resultants respectively of the several forces which are applied at the several angles as shown. These forces being in equilibrium may be represented in magnitude and direction by the sides of a triangle, as a, b, c, Fig. 123. Since the forces at C, Fig. 122, including the strains in the bars, are in equilibrium, *they* may be represented in magnitude and direction by the three sides of a triangle. In Fig. 123, draw the lines 1 and 2 from the extremities of P_1, parallel to 1 and 2 in Fig. 122, and they will meet in some point as O. The lines Oc and Oa will represent the strains in the sides 1 and 2 of Fig. 122. In a similar way, the forces 2, P_3 and 3 being in equilibrium at B, Fig. 122, we draw from O, Fig. 123, the line Ob, parallel to BA, and it must intersect cb at b, the intersection of the lines P_2 and P_3.

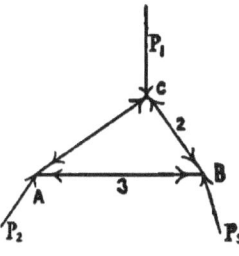

Fig. 122.

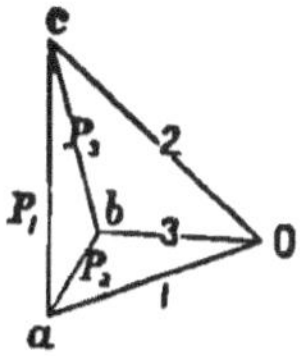

We thus see that the radial lines drawn from the point O, parallel to the sides of the triangle ABC, to the angles of the triangle which represent the external forces, represent the strains upon those sides. The same is true when the forces are applied at the angles of a polygon and are in equilibrium. Hence we have this:

THEOREM. *When the forces which are applied at the angles of a polygonal frame, are in equilibrium among themselves, we find the strains upon the several pieces which form the contour of the*

Fig. 123.

APPENDIX.

frame, by drawing radial lines from any point parallel to the sides of the frame, and cutting those lines by others which are parallel to the direction of the forces, and whose successive intersections are on the successive radial lines. The distances so cut off on the radial lines will represent the strain on the corresponding lines of the polygon.

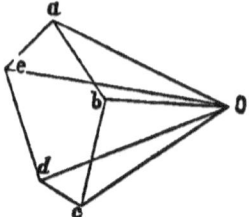

Fig. 124

If the applied forces are parallel to each other the construction becomes very simple, as shown in Fig. 40, page 55, and Fig. 41, page 56.

This method may be applied to internal trussing.

APPENDIX II.

TABLE

Of the Mechanical Properties of the Materials of Construction.

NOTE.—The capitals affixed to the numbers in this table refer to the following authorities:—

B. Barlow. Report of the Commissioners of the Navy, etc.
Be. Bevan.
Bn. Buchanan.
Br. Belidor, Arch. Hydr.
Bru. Brunel.
C. Couch.
Cl. Clark.
D. Darcel, Annales for 1858.
D. W. Daniell and Wheatstone. Report on the stone for the Houses of Parliament.
E. Eads.
F. Fairbairn.
G. Grant.
H. Hodgkinson. Report to the British Association of Science, etc.
Ha. Haswell. Eng. and Mech. Pocket-Book, 1869.
J. Journal of Franklin Institute, vol. XIX. p. 451.
K. Kirwan.

Kl. Kirkeldy.
La. Lamé.
M. Mischembroeck. Introd. ad Phil. Nat. L
Ma. Mallet.
Ml. Mitis.
Mt. Mushet.
Pa. Colonel Pasley.
R. Rondelet. L'Art de Bâtir, IV.
Ro. Roebling.
Re. Rennie. Phila. Trans., etc.
S. Styffe. On Iron and Steel.
T. Thompson.
Te. Telford.
Tr. Tredgold. Essay on the Strength of Cast Iron.
W. Watson.
Wa. Major Wade.
Wn. Wilkinson.

* Calculated from the experiments of Fairbairn and Hodgkinson.

NAMES OF MATERIALS.	Weight of one cubic ft. in lbs. s	Tenacity per sq. inch in lbs. T.	Crushing Force per square inch in lbs. C.	Modulus of Rupture. R.	Coefficient of Elasticity. E
METALS.					
Antimony—					
Cast.................	281.25	1,066 M.			
Bismuth...............	613.87	3,250 M.			
Brass—					
Cast.................	525.00	17,968 Re.	10,304 Re.		9,170,000
Wire-drawn...........	534.00				14,230,000
Copper—					
Cast.................	537.93	19,072	29,279 Re.		
Sheet................	549.06	32,184			
Wire-drawn...........	560.00	61,228			
In Bolts..............		48,000			
Iron.					
Cast Iron.					
Old Park.............				43,240 T.	18,014,400 T.
Carron, No. 2—					
Cold Blast...........	441.62	16,683 H.	106,375 H.	38,556 H.	17,270,500 H.*
Hot Blast............	440.37	13,505 H.	108,540 H.	37,503 H.	16,085,000 H.*
Carron, No. 3—					
Cold Blast...........	443.37	14,200 H.	115,442 H.	35,960 F.*	16,246,966 F.
Hot Blast............	441.00	17,715 H.	133,440 H.	42,637 F.*	17,873,100 F.

TABLE.—Continued.

Names of Materials.	Weight of one cubic ft. in lbs. s	Tenacity per sq. inch in lbs. T.	Crushing Force per square inch in lbs. C.	Modulus of Rupture, R.	Coefficient of Elasticity. E.
Iron.					
Cast Iron.					
Devon, No. 3—					
Cold Blast.............	455.93	....	....	36,288 H.*	22,907,700 H.
Hot Blast..............	451.81	29,107 H.	145,435 H.	43,497 H.*	22,473,650 H.
Buffery, No. 1—					
Cold Blast.............	442.43	17,466 H.	93,366 H.	37,503 H.*	15,381,200 H.
Hot Blast..............	437.37	13,434 H.	86,397 H.	35,316 H.*	13,730,500 H.
Joed Talon, No. 2—					
Cold Blast.............	434.06	18,855 H.	81,770 H.	33,453 F.*	14,513,500 F.
Hot Blast..............	435.50	16,676 H.	82,789 H.	33,696 H.*	14,323,500 F.
Elsicar, No. 1—					
Cold Blast.............	439.37	....	....	34,587 F.*	13,981,000 F.
Milton, No. 1—					
Hot Blast..............	436.00	....	....	29,889 F.*	11,974,500 F.
Muirkirk, No. 1—					
Cold Blast.............	444.56	....	....	36,693 F.*	14,003,550 F.
Hot Blast..............	434.56	....	....	33,850 F.*	13,294,490 F.
Morris Stirling's 2d quality..	....	25,764	119,000		
Gun Metal—					
American.............	....	....	14,000 to 34,000 Wa.	....	27,548,000 Wa.
Extra Specimens........	595.00	....	45,970 Wa.		
Steel.					
Hammered Cast Steel, from F. Krupp..................	....	91,000 S. 122,000 S.	....	....	31,359,000 S.
Tempered	....	171,000 S.			
Bessemer Steel, from Högbo, marked 10..................	....	140,945 S.	....	....	31,819,000 S.
Bessemer Steel, Eng. Mean of four Experiments........	485.37	88,415 F.	225,568 F.	....	29,215,000 F.
Naylor, Vickers & Co. Crucible Steel..................	488.70	108,099 F.	225,568 F.	....	30,278,000 F.
Mushet's Steel—					
Soft....................	492.50	93,616 F.	....	....	31,901,000 F.
Cast Steel—					
Soft....................	486.25	120,000			
Not Hardened........	....	....	193,944 Wa.		
Mean Temper.........	....	....	391,985 Wa.		
Razor Tempered........	490.00	150,000	....	....	29,000,000
Steel Wire Rope—					
Fine Wire.............	....	40,000 Bo.			
Chrome Steel............	....	195,000			
Wrought Iron.					
English..................	481.20	57,300 La.			From 22,000,000 to 28,000,000
In Bars.................. {	475.50 487.00	57,300 La.			
Hammered..............	....	67,200 Bru.			
Russian.................	....	60,480 La.			
Swedish, in bars........	....	71,680 R.			
English, in wire 1-10 inch diam................	....	80,000 Te. 96,000 Te.			
Russian, in wire; diam. 1-20 to 1-30 inch............	....	134,000 La. 203,000 La.			

APPENDIX. 247

TABLE.—*Continued.*

NAMES OF MATERIALS.	Weight of one cubic ft. in lbs.	Tenacity per sq. inch in lbs. T.	Crushing Force per square inch in lbs. C.	Modulus of Rupture. R.	Coefficient of Elasticity. E.
Wrought Iron.					
Rolled in sheets and cut crosswise...................		40,320 Ml.			
Cut lengthwise.............		31,300 Ml.			
In chains, oval links, iron ½ in. diam...............		48,160 Br.			
Wire, American............		73,600 Ha.			
Lake Superior and Iron Mountain Charcoal Bloom.		90,000 Ha.			
Missouri Iron, bar.........		47,000 J.			From 22,000,300 to 28,000,000.
Tennessee, bar, 21 exp.....		52,099 J.			
Salisbury, Ct., 40 exp......		58,000 J.			
Centre Co., Pa.. 15 exp....		58,400 J.			
Phillipsburgh Wire, Pa.					
Diam. { 0.333 in., 13 exp...		84,186 J.			
0.190 in., 5 exp...		73,888 J.			
0.156 in., 5 exp...		89,162 J.			
Mean of 188 rolled bars....		57,557 Kl.			
Mean of 167 plates lengthwise...................		50,737 Kl.			
Mean of 160 plates crosswise.		46,171 Kl.			
Low Moor, bars............		60,364 Kl.			
Swedish, forged............		{41,000 Kl. 50,000 Kl.}			
Hammered Bessemer Iron, from Hügbo.............					32,320,000 S.
Low Moor Rolled Puddled Iron...................					31,976,000 S.
Rolled Iron, Swedish, charcoal heath.............		65,000 S.			27,000,000 S.
Lead, cast, English.........	717.45	1,824 Re.			
Lead Wire.................	705.12	2,561 M.			
Silver, standard............	644.50	40,902 M.			
Tin, cast..................	455.68	5,322 M.			4,608,000 Tr.
Zinc......................	439.25				13,080,000 Tr.
STONE—NATURAL AND ARTIFICIAL.					
Granites.					
Aberdeen, blue............	164		10,914 Re.		
Cornish...................	166		6,856 Re.		
Killincy, very felspathic...			10,780 Wn.		
Mount Sorrell, granite.....	166		12,286 F.		
Sandstones.					
Caithness Pavement.......			6,493 Bn.		
Dundee Sandstone.........	158		6,630 Re.		
Derby Grit, a red, friable Sandstone...............	143		3,142 Re.		
Do. from another quarry..	156		4,345 Re.		
Limestones.					
Limestone, Magnesian (Grafton, Ill.)..............			17,000 R.		Same as W't Iron. E.

APPENDIX.

TABLE.—*Continued.*

NAMES OF MATERIALS.	Weight of one cubic ft. in lbs. s	Tenacity per sq. inch in lbs. T.	Crushing Force per square inch in lbs. C.	Modulus of Rupture, R.	Coefficient of Elasticity, E.
Limestones.					
Limestone, compact.........	162	....	7,713 Re.		
Limestone, Kerry, Listowel Quarry, Eng..............		....	18,043 Wn.		
Chalk.....................		....	501 Re.		
Other Stones.					
Alabaster (Oriental), white..	170				
Marble, statuary...........			3,216 Re.		
Do. white Italian, veined.	165	....	9,681 Re.	1,062	25,200,000 T.
Do. black Galloway......	168	....	9,219 Re.	2,664	
Portland Stone (Oolite)......	151	....	3,792 Re.		
Valentia, Kerry (slate stone).		....	10,943 Wn.		
Green Stone, from Giant's Causeway..................		....	17,220 Wn.		
Quartz Rock, Holyhead (across lamination).........		....	25,500 Ma.		
Quartz Rock (parallel to lamination)...................		....	14,000 Ma.		
Gravel......................	120				
Green Moor.................	158	....	2,010 Re.		
Artificial Stone.					
Brick, red..................	135.5	280	808 Re.		
Brick, pale red.............	130.31	300	562 Re.		
Brick, common.............		....	800 to 4,000 Ha.		
Bire Brick, Stourbridge.....		....	1,717 Re.		
Brick, Stock................		....	2,177 Ha.		
Bricks set in cement (bricks not very hard)..............		....	521 Cl.		
Brick Masonry, common.....		....	500 to 800 Ha.		
Cement, Portland, with sand.........		98 to 284 D.			
Cement, Portland, with no sand.......................		427 to 711			
Cement, Portland............		....	1,000 to 5,900 G.		
Chalk.......................	116.81	....	334 Re.		
Glass, Plate................	153.31	9,420	....		
Mortar.....................	107	50	120 to 240 Ha.		
Timber.					
Acacia, English.............	47.37	16,000 Be.	....	11,202 B.	1,152,000 B.
Alder.......................	50.00	14,186 M.	6,859 H.		
Apple Tree.................	49.56	19,500 Be.			
Ash { Ordinary state.......	43.12	} 17,207 B.	8,683 H.	12,156 B.	1,044,800 B.
{ Very dry..........	55.81		9,363 H.		
Bay Tree...................	51.37	12,396	7,158 H.		
Bean, Tonquin..............	67.51	....	....	20,886 B.	2,601,600 B.
Beech { Ordinary...........	53.37	15,764 B.	7,733 H. }	9,836 B.	1,353,600 B.
{ Very Dry..........	45.12	17,850 B.	9,363 H. }		

APPENDIX. 249

TABLE.—*Continued.*

NAMES OF MATERIALS.	Weight of one cubic ft. in lbs. S	Tenacity per sq. inch in lbs. T.	Crushing Force per square inch in lbs. C.	Modulus of Rupture. R.	Coefficient of Elasticity. E.
TIMBER.					
Birch, common	49.50	15,000	{ 4,533 H. 6,402 H. }	10,920 B.	1,562,400 B.
Birch, American	40.50		11,663 H.	9,624 B.	1,257,600 B.
Box, dry	60.00	19,891 B.	10,299 H.		
Bullet Tree (Berbice)	64.31			15,636 D.	2,610,600 B.
Cane	25.00	6,300 Be.			
Cedar, Canadian	56.81	11,400 Be.	5,674 H.		
Crab Tree	47.80		6,499 H.		
Deal—					
Christiana Middle	43.62	12,400		9,864 B.	1,672,000 B.
Norway Spruce	21.25	17,600			
English	29.37	7,000			
Red			5,748 H.		
White			6,741 H.		
Elder	43.43	10,290	8,467 H.		
Elm, seasoned	36.75	13.489 M.	10,331 H.	6,078 B.	699,840 B.
Fir—					
New England	34.56			6,612 B.	2,101,200 B.
Riga	47.06	{ 11,549 to 12,857 B. }	5,748 to 6,586 H.	6,643 B. 7,572 B.	1,328,800 860,000 B.
Hazel	53.75	18,000 Be.			
Lance Wood	63.87	24,696			
Larch—					
Green	32.62	10,220 B.	3,201 H.	4,992 B.	897,600 B.
Dry	35.00	8,900 B.	5,568 H.	6,894 B.	1,052,800 B.
Lignum-vitæ	76.25	11,800 M.			
Mahogany, Spanish	50.00	16,500	8,198 H.		
Maple, Norway	49.56	10,584			
Oak—					
English	58.37	17,300 M.	{ 4,684 to 9,509 H. }	10,032 B.	1,451,200 B.
Canadian	54.50	10,253	{ 4,231 to 9,509 H. }	10,596 B.	2,148,600 B.
Dantzic	47.24	12,780		8,742 B.	1,191,200 B.
Adriatic	62.06			8,208 B.	974,400 B.
African Middle	60.75			13,566 B.	2,284,200 B.
Pear Tree	41.31		7,518 H.		
Pine—					
Pitch	41.25	7,818 M.		9,792	1,225,600 B.
Red	41.06		5,375 H.	8,946 B.	1,840,000 B.
American Yellow	28.81		5,445 H.		1,600,000 Tr.
Plum Tree	49.06	11,351	3,657 to 9,367 H.		
Poplar	23.93	7,200	8,107 to 5,124 H.		
Teak, dry	41.06	15,000 B.	12,101 H.	14,722 B.	2,414,400 B.
Walnut	41.98	8,130 M.	6,635 H.		806,000
Willow, dry	24.87	14,000 Be.			

www.ingramcontent.com/pod-product-compliance
Lightning Source LLC
Chambersburg PA
CBHW021355230426
43666CB00006B/537